I0481909

An Introduction to Remote Viewing the FOREX. Schumann Resonance Coherence Secrets.

Scott Rauvers

Read the First 3 Chapters of this book FREE at

www.ez3dbiz.com/arvfour.html

This book is also available in Nook and Kindle Versions. Just
enter the title into a search engine online to
locate these versions.

Copyright © June 2018
All rights reserved. The Institute For Solar Studies on
Behavior and Human Health

ISBN-10: 1721182233

An Introduction to Remote Viewing the FOREX.
Schumann Resonance Coherence Secrets.

*Other great titles published by the Institute for
Solar Studies on Behavior and Human Health*

- ➢ Sunspot Secrets. Jesus as the Sun. Technology, Prophecy and Human Evolution

- ➢ A Plan to Unleash Creativity, Harness Intuition and Increase Money Flow

- ➢ The 2018 Feng Shui Planetary Prosperity Almanac and Ephemeris with Organizer

- ➢ The Complete Guide to Natural Toothache Remedies and Re-mineralization

- ➢ Solar Flares and Their Effects Upon Human Behavior and Health (revised 2017 edition)

- ➢ The Emerald Tablets: The Keys of Life and Death by Thoth the Atlantean

- ➢ Following the Sun. Using Sunspot Cycles to Change Your Destiny.

- ➢ Secrets to Creating Money Effortlessly using Lucid Dreaming (revised 2018 edition)

- ➢ The Official Guidebook of How to Make Tinctures and Alchemy Spagyric Formulas

You may preview the first 3 chapters of any of these books by visiting:

www.ez3dbiz.com/library.html

View the Associative Remote Viewing Dow Jones Project at

www.ez3dbiz.com/
dow_project_research_summary.html

An Introduction to Remote Viewing the FOREX.
Schumann Resonance Coherence Secrets.

The Solar Institute's Remote Viewing Series

Our remote viewing sessions the past 3 years involving more than 70 associative remote viewing sessions has cumulated all our data into a 4 part series of books.

CONSTELLATIONS AND REMOTE VIEWING

Book 1 - *Wormhole Theories, Sunspot Activity and Remote Viewing Stocks*. Topics Covered: Quantum Tunneling, Herbs for Remote Viewing, 13:30LST, The Star Arcturus, Cosmic Rays and Remote Viewing, Air Pressure, The Human Nervous System and Precedent Activity, Frequencies that Enhance the Results of Remote Viewing, Solar and Weather Conditions for Prime Associative Remote Viewing Sessions, Intuitive Biorhythms and Remote Viewing, Magnetic Midnight, the Ophiuchus Constellation, Mayer Waves, Moisture as a Medium for Conveying Information, The Associative Remote Viewing Procedure, Studies Involving Remote Viewing the Markets, Torsion Effects and Time, Magnetic Fields, Paramagnetic Materials, Angular Momentum and the Density of Time and much more!

REMOTE VIEWING HARDWARE AND TECHNOLOGY

Book 2 - *Associative Remote Viewing Technology. Secrets of Precognition and Intuition*. Topics Covered: Emotions as Sensors for Future Stimuli, Associative Remote Viewing and power of Expectation, The Maharishi Effect, Remote Viewing the Future of the Dow Jones, Remote Viewing Electronics / Technology, Dealing with Remote Viewing Interference, Schumann Resonance, Heart Math Coherence and Remote Viewing, Humidity as an Emotional Intensifier, Polarized Light, Finding the Ideal Remote Viewing "Sweet Spot", The Key of Time, The Quarter Moon, Neutrinos and the Nervous System, Tungsten and the Electroweak Force, Hydrocarbons, Barometric Air Pressure and Intuition, Maintaining Strong Brainwaves During Remote Viewing Sessions, Triboluminescence, The Color Yellow, Environmental Radiation and Remote Viewing, Biodynamic Gardening Phases and Remote Viewing, Photoelectrics and much more!

THE QUANTUM REALM AND REMOTE VIEWING

Book 3 – *Improve your Remote Viewing Accuracy Techniques using Quantum Microtubules*. **Topics Covered:** The Quantum Mind, Remote Viewing and Quantum Mechanics, The role Microtubules play in Remote Viewing, **Remote Viewing and Non-locality, The Hypothalamus and Remote Viewing, Gems and Minerals that Enhance Remote Viewing, Quantum Coherence, The Hippocampus, Empathy and Psychic Ability, Substances that Enhance Remote Viewing, Linoleic Acid and Quantum Mechanics, Quantum Photosynthesis, Dopamine and Remote Viewing, Transthyretin, Neurotransmitters and Remote Viewing, Lithium, Monoterpenes, The Signal to Noise Ratio and Remote Viewing, Essential Oils and Quantum Effects, Anesthetics, Taxol, The Pacific Yew Tree, Bacteria, Monoterpenes and Quantum Photosynthesis, Consciousness and Frequency, Meditation, Brainwave Rhythmus and Remote Viewing, Photons, Alternate Timelines and Parallel Universes, The Zero Point Field, The Best Moon Phases for Remote Viewing, Favorable Environments and Conditions for Remote Reviewing and much more!**

Preview the first 3 chapters of any of these editions by visiting:

www.ez3dbiz.com/library.html

An Introduction to Remote Viewing the FOREX.
Schumann Resonance Coherence Secrets.

Moving into our 3rd year of remote viewing the financial markets, this edition lists all our latest discoveries and technology. This fourth edition in our series validates our previous theories and hypothesis with published scientific studies confirming our theory that solar weather affects the health of the body, especially the heart. It also covers in detail the specific substances in essential oils that enhance remote viewing and goes into the details of why full moons enhance precognition. Standing waves are also briefly covered and how they enhance ARV sessions via the Schuman resonance. Seasonal cycles of the solar wind are also covered to help one narrow down the remote viewing sweet spot. The second part of this edition covers the new science of HeartMath and how one can use HeartMath to boost their intuition. A special chapter is devoted to cosmic rays showing how they influence HRV (Heart Rate) and can be used to enhance the success of remote viewing sessions. Specific instructions for all the 4 main HeartMath exercises are included. The Quick Coherence Technique, The Heart Lock-In Technique and The Freeze Frame Technique and The Inner-Ease Technique. This edition also explores our research into lengthening how far out into the future one can remote view by utilizing the nervous system from the timeline of minutes to days and at the end of this edition, we give tools one can use to find the best solar weather conditions for enhancing their success of remote viewing.

We at the solar institute hope you'll enjoy this next edition and utilize our discoveries and the techniques within as a pathway to opening the door to your unique gifts and talents...!

Quotes from this edition

"Our research has concluded that the heart receives intuitive information before the brain. This occurs approximately one second or more before"

"heart-rhythm measures (HRV) were especially successful in detecting pre-stimulus responses in pre-bet and post-bet segments. Both these occur prior to knowing the future outcome"

"Findings that a HRV win/loss response during post-bet and pre-bet segments is more evident during full-moons but not new moons"

"results of analysis of the participants across all trials are strong. It provides compelling evidence of practical nonlocal intuition"

The above quotes are summarized from a report titled: The study, Stability of Pre-Stimulus Intuition Response: A Repeated Measures Study Using Electrophysical Instrumentation, which was conducted by McCraty and his team at the Institute of HeartMath during the latter half of 2006. This is just one of many studies that we explore in this book and how anyone can utilize Heart Coherence to improve enhance their intuition and to supercharge their remote viewing sessions.

An Introduction to Remote Viewing the FOREX.
Schumann Resonance Coherence Secrets.

Chapters

Standing Waves and Music, Standing Waves caused by Strong Winds, Tidal Stress and Earthquakes, Volcanic Activity, Lunar Phase and Tidal Stress, Underground Magma Enhances the Distance Seiches can Travel, Standing Waves in Mountain Ranges, Microbaroms and Standing Waves, Microbaroms, Atmospheric Wind and Low Frequencies, Standing Waves and Remote Viewing. A Hypothesis, Standing Waves and Dipole Antennas, Displacement and Pressure, Types of Dipole Antennas. The Half-Wave Dipole Antenna, Inverted Vee, Standing Waves and Levitation, Solar Weather Frequencies Beneficial to Intuition.

Chapter 3. Utilizing the Full Moon to Develop One's Intuition
Page 47

Edward Leedskalin and Sweet 16, A Summary of the Solar Institute's ARV Sessions, The Moon as a PSI Amplifier, Wet Cupping Reduces the Severity of Migraines during Full Moons, Cosmic Rays and Psychic Ability.

Chapter 4. ESP Organs of the body.
Page 60

Castor Oil and Lymph Flow.

Chapter 5. Solar Weather and Its Effects upon Earth and the Moon.
Page 63

Earth's Magnetosphere and ESP, Cycles of the Sun's Solar Wind, The 2 Main Speeds of the Sun's Solar Wind, Cycles of Solar Wind Speeds, The Solar Wind, Full Moons and RetroPK, The 2 Main ARV Cycles, What does Deviation from the Elliptic

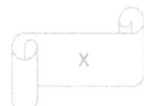

Mean?, The Solar Radiation Shielding Effect, Cosmic Rays and Computer Malfunctions, 10B Boron and Cosmic Rays, Soft Errors are Beneficial in Medical Diagnostics, Cosmic Rays and Atmospheric Changes, Seasonal Cycles of Cosmic Rays, Varying Changes in Atmospheric Pressure, Atmospheric Temperature Affects Cosmic Rays, Solar Cycles and Cosmic Ray Intensity, Long Term Solar Cycles, Cycles of Geomagnetic Storms, Cycles of Geomagnetic Activity and Moon Phase, Locating the Most Favourable Solar Weather Conditions for an ARV Session.

Why Feelings Generated by the Heart can be sensed by Others, Definition of Coherence, Positive Emotions, Sine Waves and EEG Activity, What is LF/HF?, Geomagnetic Storms and EEG Alpha Brainwaves, Coherence in Physics, Coherence in Human Physiology, Heart Coherence, Coherence Causes Increased Parasympathetic Nervous System Activity, Essential Oils that Stimulate the Parasympathetic Nervous System, Parasympathomimetic Substances, Oxytocin. The Bonding Molecule, How to Increase Oxytocin levels, Singing Increases the Social Bonding Hormone Oxytocin, Oxytocin as a Natural Fear Repellent, Herbs with oxytocic properties, Essential Oils and their Effects Upon the Heart, The Power of Limonene, Yuzu essential oil and Limonene, Limonene and White Blood Cells, Limonene in Nature, Essential Oils containing Limonene, What is a H1 Receptor?, Spices containing an abundance of Limonene, Essential Oils with an abundance of Limonene,

Foods high in Limonene, Limonene is a powerful natural Antimicrobial, Limonene and Stress, Linalool and the Parasympathetic Nervous System, Linalool Enhances the Ability to Solve Math Problems, Solar Eclipses and Bacteria, Alpha Brainwaves and Limonene, Pinene and Cymene, Eugenol, What is Eugenol?, Eugenol Synergy.

Self-Regulation Techniques, Attaining Coherence, Technology for Developing Heart / Mind Coherence, Techniques for Enhancing Coherence, The Quick Coherence Technique, The Heart Lock-In Technique, The Freeze Frame Technique, The Inner-Ease Technique, An Essential oil Blend for Generating Sustained Coherence, The Heart Lock-In Technique, Autogenic Training, Prayer and HRV.

What is Heart Intelligence?, Intuition Puts one on the Right Path, The 3 Main Types of Intuition, The Fundamental Basics of Non-local Intuition, Pre-Sentient, Meta-data Pre-sentiment Studies, The Human Heart as a Receiver for Future Information, How Coherence Enhances Intuition, Entrepreneurs and Intuition, The Full-Moon Effect and its Amplification Effects on Intuition, Pre-Stimuli and Moon Phase, The Full Moon and its Effects on Physical Endurance, What is the Step Test?, Why Self-Awareness Enhances Intuition.

Feelings of Anxiety and Irritability, What Does Polar Cap Activity Mean?, Low HRV Levels are Good for Health.

The Emerging Global Mind, RNGs, IMF Polarity, What is Positive IMF Polarity?, Future Plants of the Global Coherence Project,

The Full Moon, The Full Moon and Casino Winnings, Thunderstorms and the Full Moon, Solar Cycles and PK, Photosynthesis and Quantum Mechanics, ULF Waves and PK, Cosmic Rays and the Sun's 10.7cm Radio Flux, The Sun's 10.7cm Solar Radio Flux and Radar, Geomagnetic Activity During Full Moons Results in Less Lottery Winners, The Piezoelectric Effect, Piezo1, What are Mechanosensitive ions?.

What is RNA Expression?, Reversing Trends, Solar Activity and PSI.

The Basic Fundamentals of Initiating an Associative Remote Viewing Protocol for the FOREX and Dow Jones Markets,

Creating the Framework, Making Money on a Falling Market, Finding Favorable Solar Weather Conditions for an ARV Session, Finding the "sweet spot".

Chapter 30. Learning Self Regulation to Relieve Stress

Learning to Control Emotions Builds Resilience, The 4 Main Types of Resilience, Self-Regulation and Health, Anger and Heart Attacks, Personal Mastery and Health, Why Constant Feelings of Anger are Unhealthy, Self-Mastery is more Beneficial than a High IQ, The Nervous System, Changes in the Nervous System Attributed to Cosmic Rays, Cycles of Solar Activity, Person's Most Susceptible to Above Average Geomagnetic Disturbances, The Vagus Nerve and the Nervous System.

Chapter 31. Understanding the role HRV Plays in the Body

HRV and Resilience, Reduced HRV, Low HRV Levels are Good for Health, HRV and Neuro-Cognition, Coherence during Geomagnetic Storms, VLF, The Celestial Bridge.

An Introduction to Remote Viewing the FOREX.
Schumann Resonance Coherence Secrets.

Prologue

In my wealth building book titled: Secrets to Creating Money Effortlessly using Lucid Dreaming, I point out that accumulating a solid financial foundation need not be a stressful endeavor. As a matter of fact effortless prosperity is within anyone's grasp. This book goes into detail about attaining heart coherence, which allows one self-mastery over their emotions. This allows one better control in situations that one may feel is not within their grasp. What a powerful tool to have when one seeks effortless prosperity!

It can be a real drain on one's resources and time when stress becomes overbearing. Learning to self-regulate emotions can save a bundle of money from unnecessary health care costs and make one happier in the process. As a bonus one has stronger intuition which oftentimes helps us avoid dreadful mistakes. The ultimate goal I believe for one practicing remote viewing is the ability to gain self-mastery, which is also known as self-regulation. In the long-term it creates improved resilience. The discovery of using self-mastery to bring emotions under control is the next frontier. This will lead to a new era of human understanding and cooperation. The opportunity we face now is to learn how we can develop our intuitive potential and accelerate towards new states of being.

The Dawning of a New Age

As the influences of the constellation Aquarius continue to grow each year as we move out of the constellation Pisces, we can expect to see an increase in the interest of remote viewing. This is due to the fact that the influences of

Aquarius cause a tendency to reflect with intuition. This takes place so that mental insights are consistent with logic which has been refined so one can find clarity in information. The influence of Pisces is more of a direct form of intuition that is not always acted upon because the information received wants to be processed before it is acted upon. Hence, we are seeing a change in the way intuition is utilized and processed.

Welcome to the Fourth Edition
This latest edition on Associative Remote Viewing of the financial markets is the latest in our series. It contains numerous technical references and published papers from peer review journals as well as our own latest discoveries. The previous 3 ARV books chronicle our progress and discoveries made remote viewing the financial markets. These can be found in the following titles:

Series 1 - **Wormhole Theories, Sunspot Activity and Remote Viewing Stocks.**

Series 2 - **Remote Viewing. The Complete User's Manual on Experiencing Future Consciousness.**

Series 3 - **Improve your Remote Viewing Accuracy Techniques using Quantum Microtubules.**

This edition also includes our latest discoveries regarding the device named the **Remote Viewing Amplification** device. For instructions on how to build the device yourself with intricate details, this information will be complied and published in a separate book by the Solar Institute. You can also view a timeline of our past few years of remote viewing (beginning in 2016) the financial markets by visiting www.ez3dbiz.com/dow_project_research_summary.html

It is the goal of this edition by the Solar Institute to show how self-regulation techniques (the key to self-mastery) can be used to enhance health, intuition and remove unnecessary stress. All the techniques and discoveries discussed in this edition contain scientific references published in peer reviewed journals, confirming that the exercises, as well as influences from the sun, moon and other external environmental elements have an effect on our mental and psychological systems. This edition will also teach how one can become resistant to negative external influences while in coherence and empower themselves in the process. This edition in our series places special emphasis on the latest research showing not just how the parasympathetic nervous system responds to future events, but the underlying causes and mechanisms that make it possible as well as specific substances that can be taken to enhance its sensitivity during ARV sessions.

Much of our research early on was based on intuitive hunches and urgings, however scientific studies are now confirming our findings as fact. A special and sincere thank-you goes out to Mr. Rollin McCraty and Eckhard Etzold whose published studies now confirm much of our hypothesis.

Introduction

Years ago while living in Topanga Canyon California, I made an accidental discovery. I had invented a new non-toxic all natural special aromatherapy air freshener that contained the essential oils Sweet Orange, Ylang Ylang II, Tangerine and Lavender. Having experienced the anxiety and stress on a daily basis that accompanies sales, I began practicing the newly invented technique called HeartMath which is a simple exercise that reduces anxiety and stress, which by the way ended up working extremely well. It was during the practicing of this exercise that I found that it enhanced my accuracy of remote viewing the future position of the Dow Jones Industrial Average. How happy and relieved this made me feel! What I found even more interesting is that whenever I had sprayed the essential oil air freshener, that it enhanced my coherent abilities while practicing HeartMath. This was an unexpected bonus! So this discovery lead me to do some reserach, whereupon I discovered that specific substances in these essential oils activated / stimulated the parasympathetic nervous system, one of which is the essential oil of lavendar [1]. Further on I searched and my journey led me to studies published in Scientific Peer Review Journals showing that the human nervous system responds to future events before it takes place. This is known as pre-stimulis, pre-sentient or anticipatory reactions [2]. I go into greater detail regarding pre-stimulis responses and the human nervous system in book titled **Remote Viewing. The Complete User's Manual on Experiencing Future Consciousness.** (Series title 2).

In the end my research found studies that enhanced task

performance and cognition can be obtained simply by breathing scents of specific natural substances [3] [4]. One of the studies I came across found that inhaling the substance linalool reduced anxiety so much that it greatly increased the accuracy of people performing mathematical calculations [5] and another study found that linalool produces calm moods as well as elicits an increase in HF (high-frequency) and causes a significant decrease in heart rate [6]. This is of major importance to any of you that practice HeartMath because an increase in HF is of significant benefit to HeartMath. We shall explore the intimate details of HeartMath and how to use HF and LF power in a later chapter, including what HF means, but first let's continue with my journey seeking the answer as to why certain essential oils enhance intuition.

Being a bit of a scientist, I decided to build on the theory and formed the hypothesis that substances in essential oils may be affecting the body's nervous system. Let's explore the data for any clues.

A research study looked at the effectiveness of aromatherapy on a person's heart rate variability (HRV) and their blood pressure. The volunteers were given a blend of the essential oils of lavender (Lavandula angustifolia), lemon (Citrus limonum) and ylang ylang (Cananga odorata) which were mixed in the ratio of 2:2:1. Another group, acting as the control, were given **limonene** (35 cc) and Citral (15 cc). Please note it can be easy to confuse linalool with limonene, although they both exhibit very similar effects. All participants were told to inhale the mixtures twice daily for 3 weeks. The study found noticeable differences in the volunteer's systolic blood pressure with notable differences in their sympathetic nervous systems activity of heart rate and

and its variability (p=.047). The study concluded that this combination of essential oils is effective in lowering a person's systolic blood pressure and reducing the activity of their sympathetic nervous system [7].

In summary, these specific substances not only lower a person's blood pressure, but also reduce the activity of their sympathetic nervous system. When a person's sympathetic nervous system is reduced, their parasympathetic nervous system becomes stimulated. As we shall show in greater detail later on, a stimulated parasympathetic nervous system greatly enhances pre-stimulis responses, allowing one **greater sensitivity to detect and retrieve information from the future**. The substances linalool and limonene also act as protective shields against bad bacteria that can make the body vulnerable due to the stress obtained during an ARV session.

What Does HRV Stand for?

Biological systems in sound health exhibit patterns that can be interpreted using mathematical abstract. As a heart beats, small changes occur in-between each heart beat. Heart rate variability (HRV) measures these changes in real time and labels them IBI's (interbeat intervals). Healthy hearts do not follow a distinct pattern. The beating of a healthy heart is complex, is constantly changing and adapting to its surrounding environment. This allows the cardiovascular system to adjust itself on demand to sudden physical challenges and//or changes to homeostasis.

Reference

An Overview of Heart Rate Variability Metrics and Norms. Fred Shaffer and J. P. Ginsberg. Sept 2017.

Foods that Favourably Influence HRV

Taking fish-oil short term has been shown to influence heart rate variability. This was indicated by an enhanced vagal tone (Short-term effects of fish-oil supplementation on heart rate variability in humans: a meta-analysis of randomized controlled trials. March 2013).

Another study found that omega 3 fatty acids increased resting HRV (Effect of dietary omega-3 fatty acids on the heart rate and the heart rate variability responses to myocardial ischemia or sub-maximal exercise. George E. Billman and William S. Harris. June 2011). Hence it may be that taking omega 3's before going into coherence may help one achieve coherence faster. Further studies are necessary to see if indeed omega 3's enhance the rate at which one experiences coherence and if it does so especially during geomagnetic storms and full moons.

Further **Reading**

Omega-3 Polyunsaturated Fatty Acids and Heart Rate Variability. Jeppe Hagstrup Christensen. Nov 2011.

Effects of omega-3 fatty acids on Resting Heart Rate, Heart Rate Recovery After Exercise, and Heart Rate Variability In Men With Healed Myocardial Infarctions and Depressed Ejection Fractions. O'Keefe JH Jr. et al. April 2006.

Polyunsaturated fatty acids extend life span through the activation of autophagy. Eyleen J. O'Rourke. Et al Feb 2013.

Basil contains Linalool

The spice Basil contains an abundance of linalool. A study

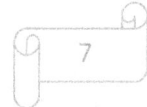

found that the major constituents in sweet basil were linalool, citral, 8-cineole, thymol, camphor, eugenol, methyl eugenol, methyl cinnamate, methyl isoeugenol and elemicine [7b]. Marotti et al., [7c] found that European basil contained methyl chavicol and linalool as the major constituents.

Limonene Protects against Free Radicals

Free radical theory states age is caused by a build-up of damage caused by reactive oxygen species (ROS) [8]. Studies have shown that **limonene** inhibits inflammatory actions by **suppressing the production of reactive oxygen species** [9]. Hence, limonene also exhibits anti-aging properties. When the body is in a relaxed state of being, it generates a healthy circulation of airflow through the lungs. Limonene can enhance this airflow. This was revealed in a study looking at whether or not limonene improves airflow in mice with asthma. The study found that after giving the mice limonene, that it greatly reduced their asthma symptoms [10].

Now that you have the overall details about how I unravelled the mystery for my enhanced intuitive abilities, I want to share with you the experiences I have had using this enhanced intuition to remote view the stock market and FOREX.

References. Introduction.

1. Lavender and the Nervous System. Peir Hossein Koulivand et al. March 2013.

2. Electrophysiology of Intuition: Pre-stimulus Responses in Group and Individual Participants Using a Roulette Paradigm. Rollin McCraty, Ph. March 2014.

3. The Effects of Linalool and Peppermint Aroma on Cognitive Performance. Kaufman, Robert

4. The Ohio State University. Department of Psychology Undergraduate Research Theses 2017. May 2017.

5. The Effects of Linalool and Peppermint Aroma on Cognitive Performance. Kaufman, Robert et al. May 2017. The Ohio State University. Department of Psychology Undergraduate Research Theses 2017.

6. Sedative effects of the jasmine tea odour and (R)-(-)-linalool, one of its major odour components, on autonomic nerve activity and mood states. Kuroda K et al. Oct 2005.

7. Effects of aromatherapy on changes in the autonomic nervous system, aortic pulse wave velocity and aortic augmentation index in patients with essential hypertension. JH Cha et al. Oct 2010.

7b. Kruger H, Wetzel SB, Zeiger B. The chemical variability of Ocimum species. J Herbs Spices Med Plants. 20029:335–44.

7c. Marotti M, Piccaglia R, Giovanelli E. Differences in essential oil composition of basil (Ocimum basilicum L.) Italian cultivars related to morphological characteristics. J Agric Food Chem. 199644:3926–3929. doi: 10.1021/jf9601067.

8. The Free Radical Theory of Aging Is Dead. Long Live the Damage Theory! Vadim N. Gladyshev. Feb 2014.

9. Limonene inhalation reduces allergic airway inflammation

in Dermatophagoides farinae-treated mice. Hirota R et al. Inhal Toxicol. 2012 May24(6):373-81. doi: 10.3109/08958378.2012.675528.

10. Limonene inhalation reduces allergic airway inflammation in Dermatophagoides farinae-treated mice. Hirota R et al. Inhal Toxicol. 2012 May24(6):373-81. doi: 10.3109/08958378.2012.675528.

Further Reading

Synchronization of Human Autonomic Nervous System Rhythms with Geomagnetic Activity in Human Subjects. Rollin McCraty et al. July 2017.

Solar and Geomagnetic Activity effects on Heart Rate Variability. Natural hazards. Dimitrova S, Angelov I, Petrova E. 201369:25–37. doi: 10.1007/s11069-013-0686-y.

The University of Boulder, Colorado in 2013 (C. Carson Smith, et al.. 2013).

An Introduction to Remote Viewing the FOREX.
Schumann Resonance Coherence Secrets.

Chapter 1. Experiences Remote Viewing the Stock Market.

Exploring the mysteries of remote viewing began for me after I read a paper published by Boulder University titled: Stock Market Prediction Using Associative Remote Viewing by Inexperienced Remote Viewers, that was published by Christopher Carson and colleagues in December of 2013. I was intrigued by this paper, not just because Boulder Colorado was where I was born, but that because students at Boulder University in Colorado, which were novice remote viewers, were able to predict the future closing position of the Dow Jones. However further research revealed that this effect was almost impossible to duplicate in further studies and that novices or first time remote viewers are blessed with what's called 'beginner's luck'. However, it is the sole aim of the remote viewing series of books to seek out ways to repeat the success of remote viewing the future position of the financial markets.

The type of remote viewing discussed in this book is Associative Remote Viewing (ARV), which means the obtaining of information from the future. Remote viewing in general is receiving information from any place no matter what the distance is in real time. For example, a person could remote view the activities of a certain individual 1,000 miles away and draw a picture of their activities as they are taking place in real time. Not much of a comforting thought for anyone who might be taking a shower! An Associative Remote Viewing Session for example can involve drawing the closing activity of a FOREX currency or Dow Jones position up to 4 days into the future. Going beyond 4 days seems to reduce the accuracy because the long term future is unstable unless it is a major earth moving event. It is only during favorable solar weather conditions (which lasts an average of 4 days) that the future is more 'set in stone' and less susceptible to changes. We shall cover why this is in greater detail later on.

Personal Changes Experienced while performing Associative Remote   Viewing

Over the years remote viewing the dow jones industrial average has led to many exciting life changes, not just for myself, but for the readers of the remote viewing series of books (www.ez3dbiz.com). One major positive change has been reduced stress and enhanced intuition. There is nobody on this earth immune to stress. Remote viewing sessions are conducted in a setting that encourages peace and relaxation during favorable solar weather conditions. Another positive benefit has been enhanced intuition. This has led to the Solar Institute publishing better much better quality articles filled with more exciting discoveries. And since the implementation of HeartMath during ARV sessions, a tremendous increase in vitality, energy and health has been experienced.

To enhance the accuracy of an associative remote viewing session, specific foods, which are generally foods that promote a strong heart and lower blood pressure, are used. This newfound energy allows one to experience more of life. A final positive benefit has been an enhanced awareness of solar weather conditions, which as we shall show later on in this book, can worsen an existing health condition, or if a person is overly stressed can cause illness, most notably a weaker immune system or increased blood pressure. The best part is having more self-control over diet, as for some people a healthy diet can be a bit of a challenge.

In summary, remote viewing is not just about gaining future information and knowledge, it is a positive transformational process that is empowering. The final benefit is much more free time. The technology has been designed to be deployed once or twice a month taking up a total of 16 hours per session. This includes solar weather forecasting, preparation materials and the actual session. If you were to do 2 ARV sessions (Associative Remote Viewing) sessions a month you would spend approximately 32 hours. The average American works 30 hours or so per week.

16 hours a month spent on ARV sessions.

120 hours a month working a standard '9 to 5' job.

Even if you didn't remote view the markets and implemented a remote viewing protocol into your work, you would still save tremendous amounts of time and energy by utilizing your mind to reduce the amount of time needed to locate valuable information or seek out more efficient ways of doing things.

Remote Viewing allows one more freedom, less stress, enhanced self-confidence and better health. It is normal to fear what one does not understand. People fear what they may become from using their gifts and talents or feel criticized for using them. This is a basic human trait. The key point here is that many people can't remote view because they cannot control their emotions during the session. The key is to learn to calm your emotions and learn to read/detect emotions from the future. Hence, once you know yourself, you become master of your destiny. Another reason some fear remote viewing or regard it as 'spooky science' is due to the fact that they are afraid or embarrassed of their true potential. These fears are due to one not having access to the real tools of self-mastery. For anyone reading this, that fear is now unfounded, as this book has all the self-mastery tools one could ever possibly need.

The Protective Barrier

We may not be able to fully access our psychic abilities at will because an invisible 'barrier' separates us from the ability. This barrier exists due to the overwhelming experiences / emotions felt when one accesses their psychic abilities. Without this barrier, information contained within the unconscious would flow freely without any type of restraint into the conscious mind. Hence, the conscious mind would not be able to take such a large 'download' of information. One way to get around this is through using specific essential oils such as limonene and linalool as described earlier. Another method is through dreams. Psychic Joseph Banks Rhine suggested that dreams are one the of most efficient and safest methods for accessing our psychic gifts. This is due to the fact that barriers of the unconscious mind are not as thick or the veil between these two worlds is thinner when one is dreaming. One of the most interesting side-effects from working with this technology is the enhancement of very lucid-vivid dreams which take place around the date(s) the ARV session is conducted.

One of the key breakthroughs that we made while researching Associative Remote Viewing was the phase of the moon. Numerous studies show that the moon affects health and emotions, however many of these studies lack the influence of solar activity because when many of these studies were conducted, we did not have SOHO and other sun space telescopes that looked at frequencies emitted by the sun. Only very recently are we beginning to understand that the effects of the moon are influenced by solar activity. This can be both good and bad, if you know what to look for. Let's now dive into the data to understand this twin interaction and how it can be used to enhance Associative Remote Viewing Sessions.

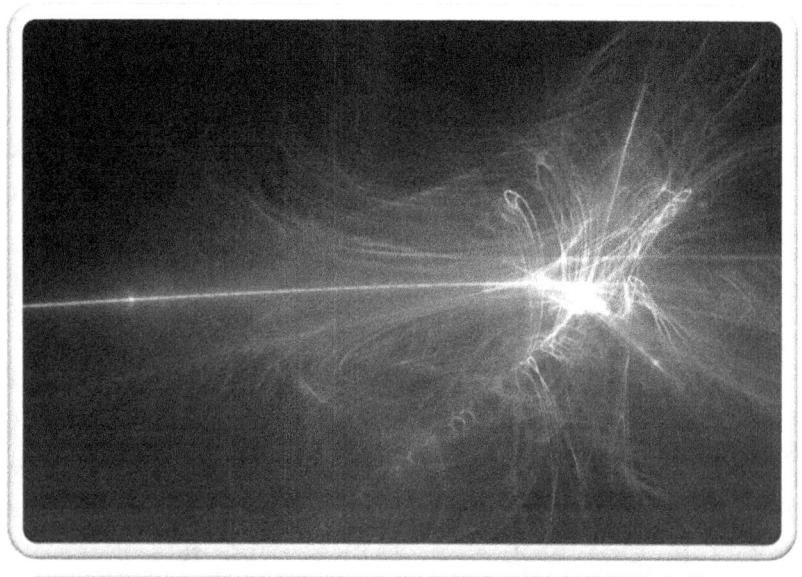

Chapter 2. Frequencies Emitted by Solar Activity and the Moon.

Every sound that we hear with our ears consists of a specific frequency. If you have ever seen the science fiction movie 'Frequency', it is about the story of a fireman's son who ends up using his deceased father's old ham radio set to speak to him decades ago while he was alive. The communications link that made this possible was that whenever a large solar flare occurred, the fireman's son was able to use the ham radio to re-connect to the past. Everything is frequency and in associative remote viewing we connect with the resonant frequencies of the future, which may really be located in another dimension that we experience as the gradual unfolding of time. It's really complicated! But yet we experience it every waking second and are completely unaware of it, just like our breathing.

Lunar Cycles and ESP

Research has proven a link between earth's geomagnetic activity and the success of ESP performance with its varying intensity (Persinger, 1989). Recent studies however suggest that there may be another lesser-known variable: phases of the lunar cycle. Numerous myths and folklore have been associated with the phases of the moon since ancient times (Guiley, 1991). In the Christian and Jewish traditions, Easter and Passover are timed according to the lunar cycle and the term "lunatic" comes from the assumption that the light of full moons made people crazy. When examining the scientific literature, looking at studies over the past few decades involving the lunar cycle and its effects on crimes, suicide attempts, traumatic hospital admissions, traffic accidents and stock market activity, scientists have found positive indications (Alonso, 1993 Geller & Shannon, 1976 Lieber & Sherin,1972). Other studies however found no effects. (Coates et al., 1989 Culver et al., 1988 Rotton & Kelly, 1985).   If we look at the dates of these studies, we see that they were made before SOHO the sun space telescope was deployed. This suggests that the level of solar activity taking place during moon phase may play a role.

Early studies looking at the lunar cycle and PSI were conducted by Andrija Puharich (1973), a neurologist working at Northwestern University where he examined a person's performance using telepathy cards. Puharich discovered that his most successful telepathy trails increased towards full moon. The success than diminished at quarter-moons and then increased again during new moons (pp. 281 – 289). Speaking from our research at the Solar Institute, the accuracy of our ARV sessions is greatly enhanced when solar wind speeds are low (350) and the moon is full. The lower solar wind speed happens to be a time when solar activity is quiet or is just starting to decline after a peak of strong activity. This low solar wind speed has also been confirmed by other independent researchers to enhance the success of remote viewing. Later on in this book we shall show a published study showing that extremely high levels of solar activity impact and 'rattle' the nervous system of the body, contributing to negative inflammation. Let's get back to the moon.......

PSI performance and lunar activity was recently addressed in a study by Eckhard Etzold (2005) of the Gesellschaft fur Anomalistik in Heidelburg, Germany. The study was known as The Fourmilab Retropsychokensis Project (**RetroPsi**). Etzold's study involved reproducing retro-PK effects on days before and after full moons. His research revealed a positive influence, with the odds against chance being approximately 1,671 to 1. Etzold then repeated the tests again 2 years later and this time observed a statistically different result during full moons, which had odds against chance of approximately 156 to 1. During his second series of tests the influence appeared to have changed direction (reversed). Etzold (2005) further found that solar activity (Solar Minimum and Maximum) was responsible for this reversal effect. In summary he concluded that retro-PK effects are modulated by complex interactions between energetic emissions from our sun and a barrier that is created by Earth's magnetic field which occurs around full moons.

Below is a quote from the study titled: A Repeated Measures Study Using Electrophysical Instrumentation, which was conducted by McCraty and the Institute of HeartMath's Mike Atkinson during the latter half of 2006.

"**The findings that HRV win/loss response during both pre-bet and post-bet segments, during the full-moon phase, but not the new-moon phase, are very intriguing and worthy of additional study.**" (McCraty et al., 2004a, 2004b).

The Magnetosphere

A large magnetic field envelopes our earth which consists of a series of layers which deflects solar radiation. This barrier is called the magnetosphere (Lyon, 2000). If we were to observe this barrier from outer space using special equipment, we would see a deflection of cosmic radiation around our Earth consisting of waves, much like the wake of a boat leaves upon the surface of water as it passes by. The shape of this wake is a tear drop type shape that bulges at the front as the radiation diverges and sweeps around the Earth. This then narrows down creating behind it a tail of gas which streams out like a comet's tail. Hence the name 'magnetotail'. As our moon orbits earth, it enters and passes through the magnetotail due to the angle of its varying orbit. During a full moon, it moves deeper into the magnetotail.

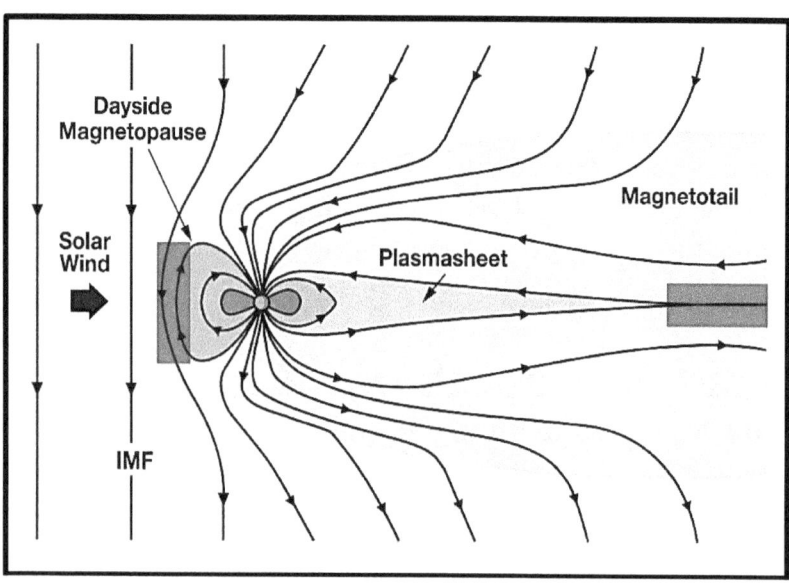

Shown above is earth as the small dot in the center. As can be seen in the next image, when the moon is full, it is behind the earth and deep inside earth's magnetotail.

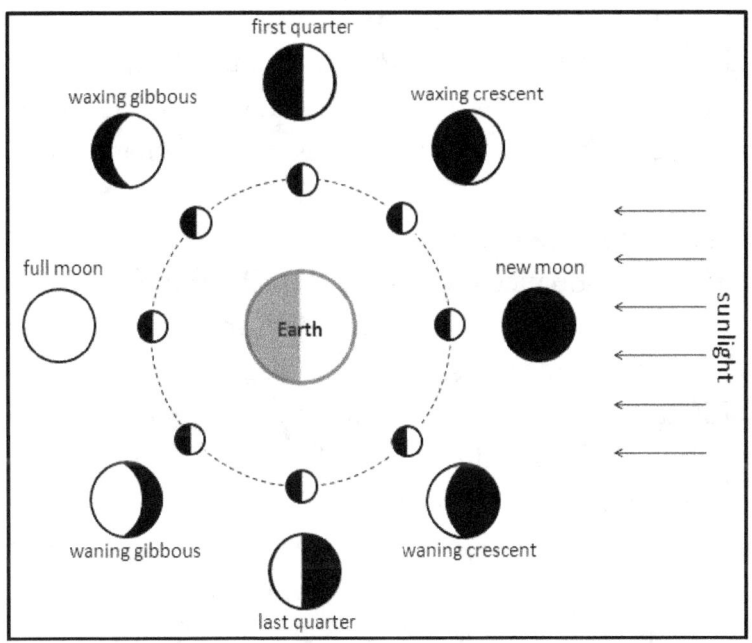

Etzold's (2005) studies have led to the hypothesis that retro-PK effects could be modulated by interactions of our moon with earth's magnetosphere when it is full, and that the moon is emitting electromagnetic waves that occur in the ultra-low frequency (ULF) spectrum as it interacts with earth's magnetosphere and the sun's solar wind. The solar wind may be amplifying or reducing this effect. Hence these varying ULF waves may be affecting human behaviour.

The Sun's 10.7cm Solar Radio Flux. The Keyhole to the Doorway of the Sun.

What is the sun's 10.7cm Radio Flux?
This index measures the amount of "solar noise" that is emitted by the sun. Hence this may be a form of noise that is conductive to enhanced mental focus and concentration.

Thunderstorms and the Full Moon
The Schuman resonance is caused by lightening, which usually occurs during thunderstorms (Application of the Schumann resonance spectral decomposition in characterizing the main African thunderstorm center. Michal Dyrda et al. Oct 2014) and during full moons thunderstorms are more frequent, especially the following two days after the full moon and that the increase may be due to earth's magnetotail (Relationship between Thunderstorm Frequency and Lunar Phase and Declination. 20 September 1970).

More Cosmic Rays Occur during Solar Eclipses and the Full Moon
Volodichev et al. (1991) observed atmospheric radioactivity, which is the intensity burst of thermal neutrons, took place during the solar eclipse of 22 July 1990. Volodichev also found thermal neutron enhancement occurred during new and full moons, including the days close to them. They attributed the enhancement to crossings of lunar tidal waves that took place over their observation site which caused deformations of cracks in Earth's crust. This in turn released trapped radioactive gases in the form of Radon into earth's atmosphere. Alpha particles that generated by Radon interact

with earth's crust and its surrounding air, creating increased neutron splashes (Volodichev et al., 1987, 1991, 1997; Antonova et al., 2007).

Dorman and Shatashvili (1961) observed during full moons that the secondary cosmic ray flux (SCR) increased, however during new moons, a decrease took place. Dorman and Shatashvili (1961) explained this increase/decrease of neutron flux during full/new moons as resulting from geomagnetic rigidity variation caused by lunar tides in earth's magnetospheric plasma (Confirmation of secondary cosmic ray flux enhancement during the total lunar eclipse of 10 December 2011. Anil Raghav et al. Oct 2013).

The study also discovered that a diurnal variation of the secondary cosmic ray flux (SCR) exists, showing a minimum during the afternoon and a **maximum during the night**. The decrease begins after sunrise which is influenced by decreasing/increasing trends in humidity/temperature. The increasing trend begins just before sunset and is influenced by increasing/decreasing trends in the humidity/temperature. The SCR flux corresponding with humidity and temperature was also found to occur in studies conducted by Raghav et al. (2013).

Further Reading
Unexpected enhancement in secondary cosmic ray flux during the total lunar eclipse of December 2011. Anil Raghav et al. Dec 2012.

Besides emitting vast amounts of light, our sun also emits numerous frequencies (Etzold. 2000 pgs 157, 161, 171). Some of these frequencies we shall cover in greater detail later on, however one specific solar frequency is the sun's 10.7cm solar radio flux. This frequency can greatly enhance the success of associative remote viewing sessions. If you were to review all the published remote viewing studies, you would see that trying to remote view numbers of future events is almost impossible. For example trying to remote view the number of a horse that will win a race is much harder than trying to remote view the colour of the jersey the jockey is wearing as he crosses the finish line. The reason we had ended up moving from remote viewing the Dow Jones to the FOREX was due to the fact that we can see the graph activity of a currency much more clearly during our sessions, rather than the actual number(s).

For example it is much, much easier to draw a pictorial graph of where the FOREX market will close in the future compared to the actual number. One very interesting thing stands out though, whenever the sun's 10.7cm solar radio flux had been rising, especially consecutively for the past few days, when we had been remote viewing the future position of the dow jones, we were able to get numerical data or get very close to the actual closing number that the dow was going to close. In other words, if you are seeking actual numbers for future data, do the ARV session when the sun's 10.7cm solar radio flux is rising or has been stronger for a number of days and you will find your accuracy will be much greater. Later on we shall show a scientific study that found that people had overall feelings of greater well-being when the sun's 10.7cm solar radio flux was increasing. This suggests a collective unconscious effect taking place during remote viewing which we shall also explore in greater detail later on. Now let's get back to the mysteries of earth's magnetotail.

The magnetotail appears to be playing a major role in ARV sessions as the moon moves deeper into it when full. Let's find out why.

Magnetotail Frequencies caused by the Moon's Orbit

As the moon interacts with earth's magnetosphere / tail, it becomes overlapped with layers of energy present in earth's magnetosphere. These overlapping waves of frequency generally consist of ultra-low frequencies (ULF) [1]. It may be that when solar activity is above average, it creates a shielding type effect, cutting the viewer(s) off from these overlapping waves. The frequency range of these overlapping waves is in the ULF (ultra-low frequency) range [1]. These frequencies have already been scientifically proven to affect human behaviour (Wilson et al. 1990) due to their ability to modulate   brainwaves.

The Solar Wind and its Interaction with Earth's Magnetosphere

The frequencies generated by the interaction of the solar wind and magnetosphere specifically fall in the 1mHz to 10Hz range (Stellmacher. 1998. Schubert, Sonett, Smith, Colburn and Schwartz. 1975 p. 279). This forms transversal and standing waves between 1mHz and 10Hz (Stellmacher, 1998). The effects of these frequencies may be causing amplification effects when a person is in heart coherence during quiet solar and geomagnetic conditions. This may lead to a period where reduced interference during remote viewing takes place. Hence generating or being in an environment where 1mHz 10Hz frequencies reside may be beneficial during RetroPK / Associative Remote Viewing Sessions. The frequency generated by the human heart while in coherence is approximately 0.1 hertz (10 seconds).

Further Reading
Toroidal Standing Waves Excited by a Storm Sudden Commencement: June 1990.

10Hz and Reactions
A study found that when 10 Hz transcranial alternating current stimulation was applied over a person's posterior cortex during visual tasks, that it caused reliable increases in their EEG alpha power. Hence it prevented deteriorations in visual performance, exerting a stabilizing effect on the person's visual attention (The Effects of 10 Hz Transcranial Alternating Current Stimulation on Audiovisual Task Switching. Clayton MS, Yeung N and Cohen Kadosh R. Feb 2018).

Standing Waves
Unlike what it sounds like, standing waves are not waves that 'stand around'

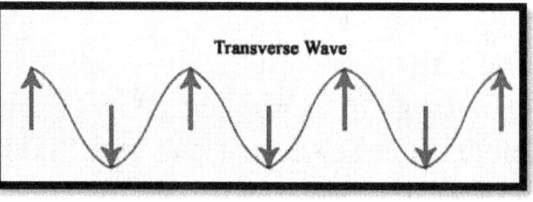

Transverse Wave

waiting for something to happen!! Standing waves, are waves most musical instruments create while making their sounds and represent the lowest energy vibrational modes of an object as it emits sound [2]. The following images show examples of transverse and standing waves. The above picture shows traverse waves.

Transverse waves occur when particle motion is perpendicular in which the direction the waves are travelling. Examples include electromagnetic waves or waves on a string. And below is an example of standing waves using a rope tied to a solid object.

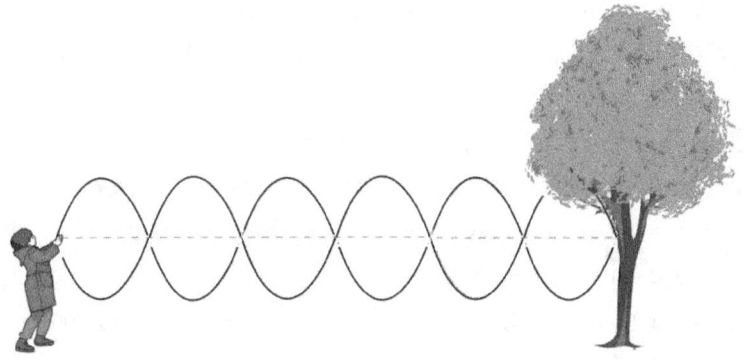

Standing waves are the result of interference of dual identical waves that travel in opposite directions.

Holograms and Standing Waves
In holograms, standing waves occur when an object wave meets a reference wave, resulting in a pattern of interference. When this is photographed, it is called a hologram.

Standing Waves and Music
The majority of sounds that we hear on a daily basis are not standing waves. They are waves that act like a pebble dropped into a pond moving outwards. The sounds we hear from musical instruments are mostly composed of standing waves. The sound waves are bounced around (reflection) or are bent (refraction). It just takes two or more surfaces to cause this effect. Musical instruments do this by producing pitches by trapping sound waves. Musical instruments take

advantage of what's called 'constructive interference' which can occur from stretched strings. This causes the waves to change phase upon reflection from a fixed end. Under these conditions, the string vibrates in regions or segments and become "standing waves". Transverse waves occur mostly on strings and standing waves in air columns or water. These standing waves in air columns form nodes and antinodes.

Standing Waves caused by Strong Winds

When standing waves occur on water they are called seiches. Watching bathwater slosh from one end to the other and back is an example of standing waves. The seiche exists as a stationary or standing wave characterized by points that appear to be standing still. The key requirement for standing waves in water is that the waves must be partially enclosed, allowing for a compressed region of energy. This enables the waves to be reflected back and forth from each other. This is why standing waves occur most often in harbors, lakes or bays. Like tsunamis, seiches can be caused by earthquakes or landslides. They occur most often during rapid changes in atmospheric pressure and strong winds as water is pushed from one end of a body of water to the other. When strong winds begin calming down and the pressure ceases, the raised water that was generated, subsides and creates the 'sloshing' motion.

Seiches can oscillate for hours or even for days, ceasing only when gravity and friction smooth them out. Seiches can be mistaken for regular tidal activity, but can be very large and dangerous. The Great Lakes in North America have the phenomena known as "the slosh." Lake Erie produces the largest and most seiches due to its shallowness and orientation. They occur when strong winds blow southwest to northeast which affects both Buffalo, NY and Toledo, OH- cities on opposite shores of the lake.

Earthquakes will generate seiches thousands of miles from their epicenter. For example, the great (magnitude 9.2) Alaskan earthquake in 1964 caused swimming pools to slosh as far away as Puerto Rico. It also triggered large seiches in bayous along the U.S. gulf coast and bodies of water in Sweden and Scotland responded to the Lisbon earthquake of 1755 with seiches up to 6 feet occuring in Norway after the Tōhoku earthquake that took place in 2011.

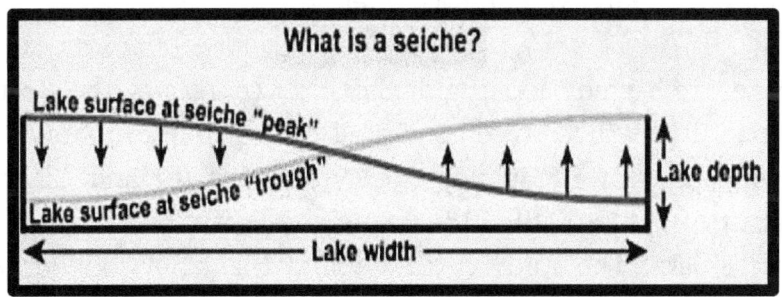

People sometimes question why a calm lake with no wind has waves rolling on the shorelines. This is because of atmospheric pressure that causes seiches which can be occuring up to hundreds of miles away.

Tidal Stress and Earthquakes

Satoshi Ide and colleagues at the University of Tokyo investigated earthquake records covering Japan, California and the entire globe. They discoverd that for 15 days leading up to each earthquake that the largest quakes that hit Chile and Tohoku-Oki occurred near maximum tidal strain (or during full and new moons). These happen to both be times when the Sun, Moon and Earth align. Further studies by his team looked at more than 10,000 earthquakes of magnitude 5.5 and above and found that earthquakes that occured during high tidal stress periods were more likely to grow to magnitude 8 or above (Tides and earthquakes. Alexandra Witze. Sept 2016).

Summary

If earthquakes are indeed more intense during new and full moons, could standing waves somehow be part of the answer? If so, this shows that standing waves are a powerful and yet untapped source of potential energy.

A research study looking at Seiches occuring at the Port of Rotterdam which were generated in the Southern North Sea found that low-frequency energy between 0.1 and 2.0 mHz occurred at sea BEFORE a seiche event took place in Rotterdam. The study further found that after looking at 6 years of weather and 51 seiche events that they coincided with the passage of low-pressure weather systems with some exhibiting sharp cold fronts (ana or classical) [3]. Also others included more diffuse cold fronts such as kata or split.

Because a tidal range can be amplified due to resonance, it allows the system to store vibrational energy. As an example the Bay of Fundy in Nova Scotia has tides over 50 feet.

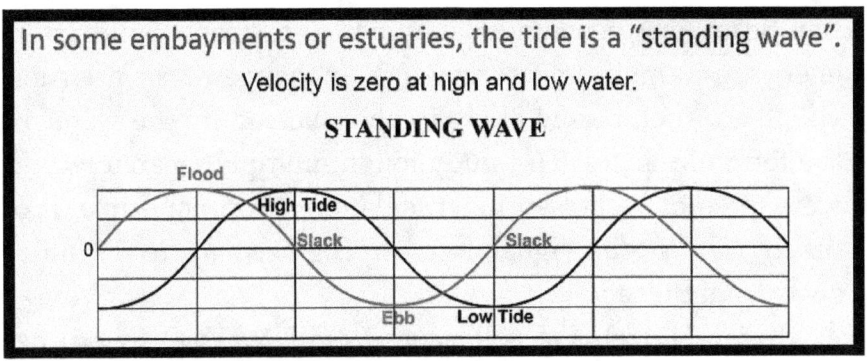

Volcanic Activity, Lunar Phase and Tidal Stress

A study looked at how fortnightly tides affected the Ruapehu volcano in New Zealand during the years 2004 to 2016. The study found that a 1-year correlation existed which found an increase in activity ~3 months before the 2007 phreatic eruption of Ruapehu. The study concluded that the volcano is sensitive to fortnightly tides and that real time monitoring of seismic sensitivity and the lunar cycle may help detect the clogging that occurs in active volcanic vents, allowing better forecasting methods (Sensitivity to lunar cycles prior to the 2007 eruption of Ruapehu volcano. Társilo Girona,. et al. Jan 2018).

Underground Magma Enhances the Distance Seiches can Travel

A research study conducted in Yellowstone National Park found that a strain signal existed, with a 78-minute frequency, originated from an seiche only an inch or two tall within Yellowstone Lake. The seiche may have been triggered by a change in barometric air pressure or high winds, which can take a few days to die down. This is a significant finding because it indicates a strong build up of energy resulting from strong winds or changes in air pressure, which are then passed along to and released by the water in the form of seiches. The study found multiple separate waves were present, representing oscillations from multiple lake basins. The seiche signals were recognized up to 19 miles away from the lake.

Karen Luttrell and colleagues from UNAVCO stated that magma beneath the ground at Yellowstone causes the seiche signals to travel further than it normally would compared to magma-free conditions. This is because the magma causes elastic-like behavior (viscoelastic behavior). These results are consistent with other studies (Smith et al., 2009).

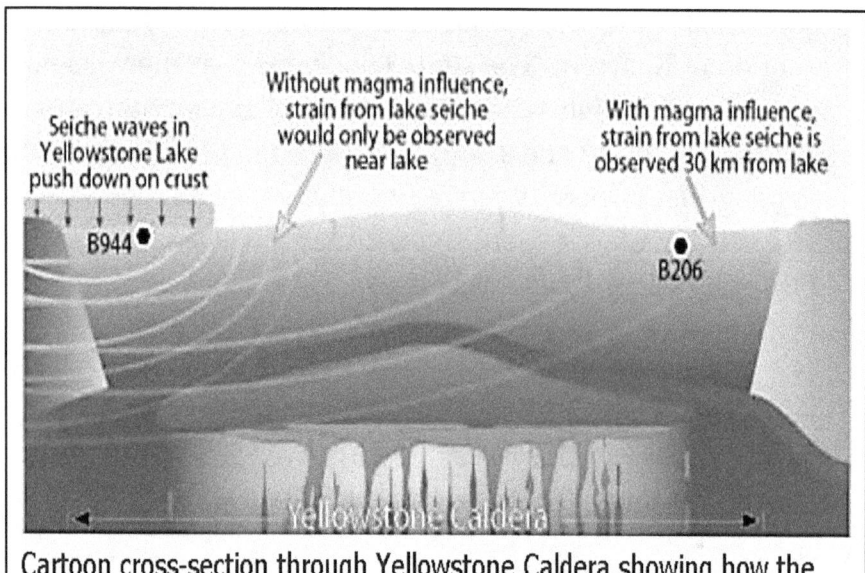

Cartoon cross-section through Yellowstone Caldera showing how the magma beneath the surface makes the Yellowstone Lake seiche detectable very far from lake. (click for larger image)

Standing Waves in Mountain Ranges

As cool air collects in a basin in a leeward valley, surges of winds occurring above the upper slopes of the valley create standing waves. This has been proven in lab experiments by Cuningham and Bedard (1993).

Mountain standing waves are caused by air being forced to rise up windward sides of mountains. Next it sinks down the leeward side. This develops into a series of standing waves downstream which may extend for hundreds of kilometers over land, open water and clear skies. If the air is sufficiently moist, the crests of these waves may be witnessed as **lenticular clouds**. Mountain waves can extend into the stratosphere and are more pronounced as height increases. Some pilots have reported these standing waves over mountains as high as 60,000 feet [4]. This may be what may have been responsible for the time-slips that have been reported by pilots which I go into more detail in my other book Improve your Remote Viewing Accuracy Techniques using Quantum Microtubules.

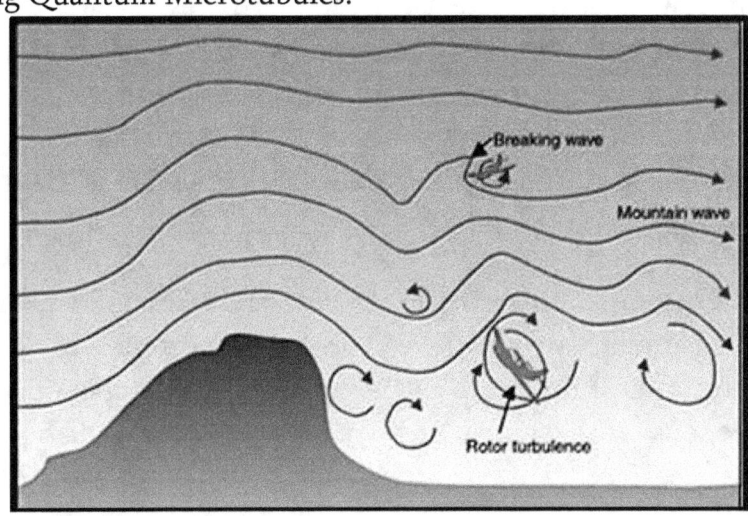

We believe the reason ARV sessions were so accurate in westward Topanga Canyon California close to the beach was because the strong winds that blew down through the valley towards the sea entered the valley creating a region of standing   waves.

Microbaroms   and   Standing   Waves

Microbaroms are a class of atmospheric infrasonic waves caused by marine storms which take place from non-linear interactions of the ocean's waves with the atmosphere. Because they are not readily absorbed by the atmosphere, their low frequency allows them to travel thousands of kilometers through earth's atmosphere and are easily detected with instruments. For example American seismologists Beno Gutenberg and Hugo Benioff at the California Institute of Technology at Pasadena detected these frequencies using a wooden box and low-frequency loudspeaker mounted on the top of the box. The microseisms were similar to seismograph activity and they correctly hypothesized the signals were from low pressure systems in the Northwest Pacific Ocean. Microbaroms can be produced by standing waves that come from two storms or if ocean swells are reflected at the shoreline. Waves at a frequency of 10-second periods are numerous in the open oceans. Microbaroms exist as low-level atmospheric infrasound between 0.1 and 0.5 Hz.

Microbaroms, Atmospheric Wind and Low Frequencies

Microbaroms are believed to come from nonlinear interactions of the oceans waves that are traveling in opposite directions with similar frequencies (Acoustic radiation by ocean surface waves. S. Arendt and D. Fritts. 2000).

Microbaroms exhibit coherent radio signals in the 0.1 to 0.5 Hz range. They can be observed anywhere on earth and are related to strong ocean wave and storm activity (Infrasonic observations of open ocean swells in the Pacific: Deciphering the song of the sea, M. Willis et al. 2004). Studies have shown that the signals caused by microbaroms could depend upon atmospheric wind conditions (On using ocean swells for continuous infrasonic measurements of winds in the lower, middle, and upper atmosphere. M. Garcés. et al. 2004). Because of their ability to travel such long distances without little interference it may be possible to utilize them as a communications source by encoding frequencies within them.

Standing Waves and Remote Viewing. A Hypothesis
During favorable solar weather conditions, the intensity of the Schuman resonance is stronger. This allows for the standing waves that occur while the full moon is in earth's magnetotail to become amplified. This allows for a clearer and more intense standing wave cavity to be generated in earth's ionosphere. Above average strong winds that occur on the leeward sides of mountains that generate standing waves may merge / connect with the ionospheric standing waves generated by the Schuman resonance, allowing a clear connection of information to be received during an ARV session. Structures or environments that naturally generate standing waves may further amplify or contain / capture the resonance of these enhanced standing waves.

As some of you know that when I perform the ARV sessions they are conducted in an old solid concrete WW2 bunker. The entry into the WW2 bunker consists of a square hatch approximate 4 feet in diameter that leads vertically downwards into a square shaped room measuring approximately 40 feet by 40 feet. As strong winds blow over the open hatch, it may be generating standing waves within. The hatch has no door and is always open.

The other region our ARV sessions were conducted was in lower Topanga Canyon, California, a region known for having extremely strong winds due to its location being in a valley that is near the ocean. The shape of the valley may be creating standing waves. It is interesting to note that a moving car generates standing waves that are generated by low frequency noise from its window being open [5].

I mention in my other book Improve your Remote Viewing Accuracy Techniques using Quantum Microtubules that microtubules, which are composed of microscopic tubular structures present in numbers in the cytoplasm of cells, may be related to the quantum effects experienced during remote viewing. It may be that standing waves are occurring in these structures. Standing waves have been found to occur in the brain's neurons.

A research study hypothesized that neurons may be related to standing waves by looking at activity between the brain's neurons. In a standing wave, the activity involves fixed sets of neurons whose responses follow the same time course [6] [7] [8]. Standing waves have so much repetitive / kinetic power that they have been shown to be able to drive a robot. A research paper describes a kinematic model using a piezoelectric miniature robot with legs that is powered by standing waves [9].

Standing   Waves   and   Dipole   Antennas
Standing waves occur on a half-wave dipole antenna which are caused by a sinusoidal voltage VO coming from a radio transmitter in its resonant frequency. This causes waves of voltage and current to reflect back and forth between the ends of the rods which cause interferrence leading to standing waves. The following image shows how this takes place.

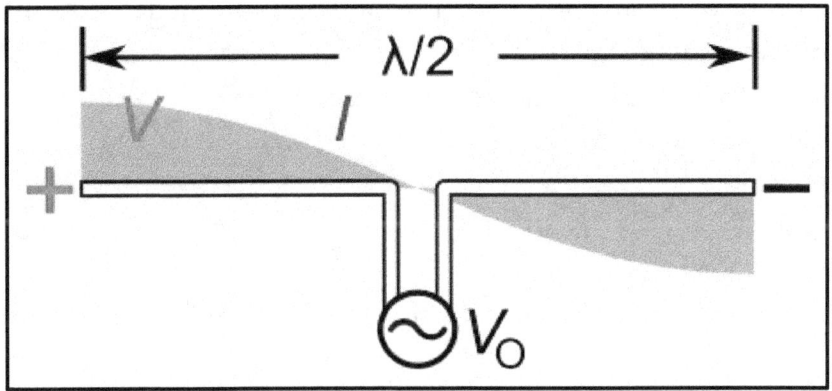

Dipole Antennas in Chlorophyll
We have found that using a bowl of algae water, which contains natural Chlorophyll enhances our ARV results. We attribute this to the quantum effects exhibited by Chlorophyll which we cover in greater detail in Improve your Remote Viewing Accuracy Techniques using Quantum Microtubules. Research studies have found that Chlorophyll has light-harvesting antenna systems that create a rapid transfer of energy. The mechanism for this energy transfer is caused by resonance energy transfer using a dipole-induced-dipole process. This was initially described theoretically by Förster. In nature, three primary antenna systems are utilized. 1- The

distance that separates the donor and acceptor chromophores. 2 - The orientations of the chromophores. 3 - The spectral overlap between donor and the acceptor chromophores [10]. Because standing waves emit so much energy, this may mean that Chlorophyll has potential for yet to be discovered energy.

Further   **Reading**

Chlorophyll transition dipole moment orientations and pathways for flow of excitation energy among the chlorophylls of the major plant antenna, LHCII. Iseri E and Gülen D. Sept 2001.

Dipole Strengths in the Chlorophyll. Robert S. Knox and Bryan Q. Springz. Department of Physics and Astronomy, University of Rochester, Rochester, NY. February 2003.

The orientations of core antenna chlorophylls in photosystem II are optimized to maximize the quantum yield of photosynthesis. Richard Cogdell. Author links open overlay panel. Sergei Vasil'ev. March 2004.

Light Absorption and Energy Transfer in the Antenna Complexes of Photosynthetic Organisms. Tihana Mirkovic.

Dipole Antennas are used in Television antennas where they are commonly referred to as "rabbit ears".

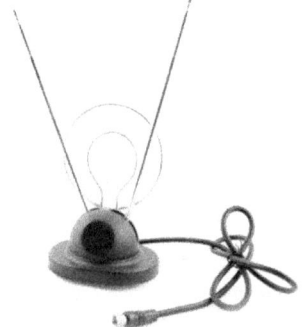

A research paper published by Mr. Jerry Stuger hypothesis that a link exists between memories and standing waves which may influence memory function and the processing of information in the brain [11]. Another research study looked at standing waves that occur near the moon [12]. Hence during full moons a synergy may exist between the standing waves near the moon and earth's magnetail.

Summary

A standing wave is an applied frequency to the resonant mode of an extended vibrating object caused by constructive interference of two waves that are travelling in opposite directions inside a medium.

Displacement and Pressure

Standing waves can also be visualized in terms of pressure variations inside a fixed region of space. Nodes for displacement act as antinodes for pressure and vice versa. When air is constrained to a node, the air is squeezed toward that point and expands away, causing the pressure variation to occur at maximum. This is why the mouthpiece end of a wind instrument acts as a node for resonances. As an example, a clarinet is acoustically a closed-end cylindrical air column due to its mouthpiece end acting as a pressure antinode. The optical cavities used by lasers uses a pair of facing mirrors. This causes a gain to take place in the cavity which causes the light to become **coherent**. This excites standing waves of light in the cavity. Also a standing wave can be produced in the air by an electrical transformer when its frequency hums at 60.00 Hz.

It may be the reason ARV session accuracy is greatly

enhanced during peaks of barometric air pressure is due to a synergy taking place between the standing waves generated in earth's magnetosphere while the full moon is immersed within it.

Types of Dipole Antennas . The Half-Wave Dipole Antenna. This is a half-wave dipole antenna which produces a maximum gain for a narrow range of variable frequencies. The dimensions of a dipole are 1/4 wavelength to 1/2 wavelength above the ground level for long-range. For extra distance, the antenna should be 1/8 wavelength and 1/4 wavelength above the ground. This same feature also applies to the sloping Vee and inverted Vee antennas.

Inverted   Vee.
The inverted Vee is also called the drooping dipole which is like a dipole but uses only just a single support in the center. It too is used for a specific frequency. Due to its inclined sides it produces a combination of vertical and horizontal radiation. It has vertical frequencies coming off the ends and horizontal frequencies coming off broadside to the antenna. All construction details for a dipole antenna apply for an inverted Vee. The inverted Vee contains less gain than a dipole antenna, however the use of needing just a single support makes this the preferred antenna in some situations.

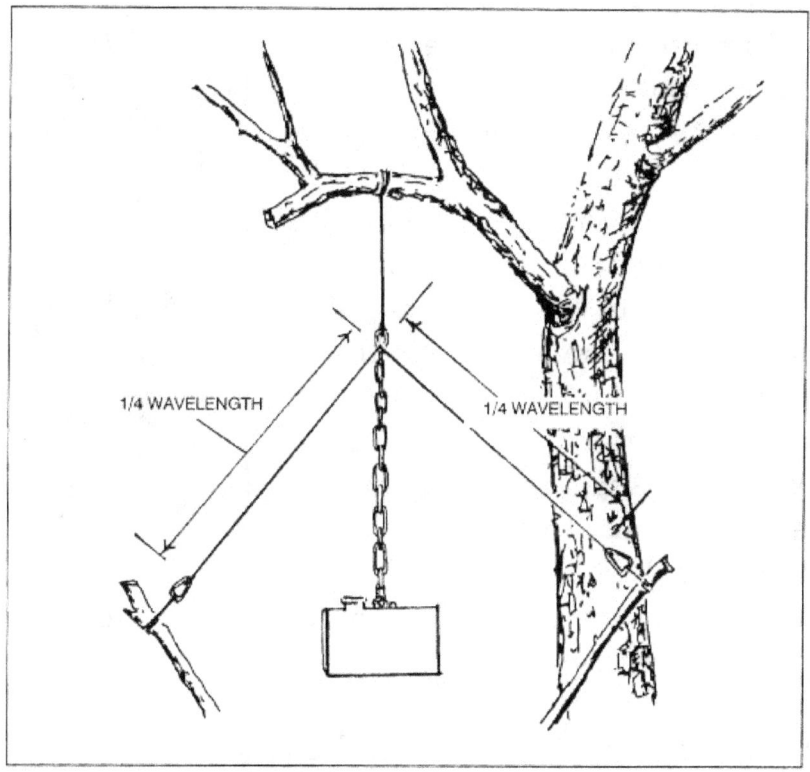

Figure D-10. Inverted Vee antenna.

When it is used as a feed line for dipole antennas, the operator connects each of the two insulated wires of the wire to a separate leg of the dipole. On the radio receiver one wire is connected to the center connector of the radio's antenna terminal with the second wire being connected to a screw on the antenna case. The Vee antenna consists of dual inverted Vee dipoles that are positioned at right angles. At the center is a foam-electric center pole uased as its coaxial.

Standing Waves and Levitation

During 2014 scientists in Japan at the University of Tokyo used four common audio speakers through which they played inaudible high frequency sound waves which intersected with one another inside a confined space. This intersection of the waves created "standing" waves.

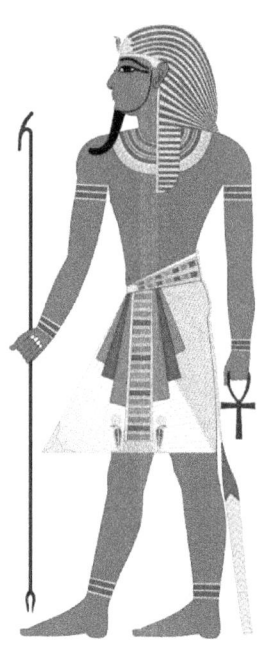

The scientists were able to use the standing waves to levitate a small screw and even direct it from side-to-side. Ancient Egyptian hieroglyphics display a metal rod known as the 'Was Scepter' (pictured). Only the priests and pharaohs were allowed to carry it. The bottom looks like a tuning fork.

Solar Weather Frequencies Beneficial to Intuition

These frequencies change depending on the amount of solar activity taking place. During full moons when earth's geomagnetic field is quiet and solar activity is low, the amplitude of these frequencies may increase (Smith, Schubert, Sonett, Colburn & Schwartz 1975 p. 279). Hence earth's moon may be exerting a type of tuning fork type effect due to the transverse / standing waves with regard to these frequencies while immersed in earth's magnetotail (Hood & Schubert, Schubert et al 1975).

The transfer function amplitude (voltage amplitude at output as a function of frequency of a amplitude wave applied to the input) when the moon is in the plasma sheet when full is approximately 1.7 to 2 for frequencies ranging between 2 and 6 mHz. This decreases to 0.7 to 1.2 transfer function amplitude when the moon is in the lobes of the magnetotail [1]. These ULF effects with the moon moving through earth's magnetotail were explored in the 1970s during the Apollo Moon Program however unfortunately these studies were discontinued [1].

Looking at the shape of a peak full moon RetroPK for-the-record data (Etzold, 2000, p. 157) the effects take place prior to the day of the full moon and then decrease after the full moon. This is when the moon crosses the high-temperature low-density plasma in the plasma sheet in the tail lobes of the magnetotail (Lichtenstein & Schubert, 1976). Because earth is rotating around our sun at such a high speed, the magnetotail is not directed straightforwardly like earth's shadow on earth's night side, but is instead asymmetrically curved in the direction of moon as it approaches us (Tsyganenko, 2002b). Hence, the time interval between the moon entering earth's magnetotail and a full moon is longer than the time interval between a full moon and the moon leaving earth's magnetotail.

Summary

ULF frequencies in the range of 1-10mHz may explain RetroPK performance taking place during full moons. This frequency may be negatively impacted during stronger solar activity due to higher frequencies being emitted by our sun. Hence, an artificially generated field wave in the 1-10 mHz range may enhance RetroPK experiments, although further research is needed to confirm this hypothesis. As the moon moves deeper into earth's magnetotail as it approaches full, standing and transverse waves are possibly being generated having an effect upon life on earth.

As covered earlier, transverse / standing waves would occur most often in enclosed cavities where strings are stretched / pulled such as musical instruments. As the moon enters earth's magnetotail, it is entering a cavity which is stretched causing standing waves. It is interesting to note here that the research discoveries on time by Dr. Kozyrev discovered that the greatest effects on the changes of time would always occur in the material Tungsten, which is one of the most flexible materials, whose metal is used to make helicopter blades (high modus of elasticity). It is also interesting to note the stretching effect is seen in the science fiction time travel movies MIB 3 and Timecop as they enter the time stream and are on their way into the past. Our ARV Device utilizes a lot of the material Tungsten., and the book Improve your Remote Viewing Accuracy Techniques using Quantum Microtubules goes into greater detail on how Tungsten is used to construct the ARV amplification device.

The primary purpose of this edition is to show the latest peer reviewed scientific technical data and use it as a reference. This has confirmed our hypothesis on our Solar Weather and its effect on health and precognition spanning the last 10 years linking the success of ARV sessions and healing / anti-aging and its relationship to Solar, Cosmic and Lunar Activity, more of which we shall cover in greater detail later on. Now that we now have a much better understanding of the mechanisms and environmental conditions responsible for enhancing one's intuition, let's see how we can use the power of the moon to enhance our intuition.

References. Chapter 2.

1. Solar Periodic full moon effect in the fourmilab retropsychokensis project. Eckhard Etzold.

2. 13-3 harmonics Understand standing wave and harmonics. Mina Crumble.

3. Generation of seiches by cold fronts over the southern North Sea. M. P. C. de Jong et al. Environmental Fluid Mechanics Section, Delft University of Technology, Delft, Netherlands. January 2003.

4. Mountain wave turbulence Mountain wave and associated turbulence. Australian Government transportation safety bureau.

5. Potential health effects of standing waves generated by low frequency noise. Ziaran S.

6. Standing waves and traveling waves distinguish two circuits in visual cortex. Andrea Benucci. Et al. July 2007.

7. Propagating Waves of Activity in the Neocortex: What They Are, What They Do. Jian-Young Wu et al. Oct 2008.

8. Standing Waves and Traveling Waves Distinguish Two Circuits in Visual Cortex Author links open overlay panel. Andrea   Benucci   et   al   .   June   2007.

9. Locomotion Study of a Standing Wave Driven Piezoelectric Miniature Robot for Bi-Directional Motion. Hassan   Hussein   Haririet   al.

10. How nature designs light-harvesting antenna systems: design principles and functional realization in chlorophototrophic prokaryotes Donald A Bryant and Daniel P   Canniffe.   Jan   2018.

10b. United States Army Field Manual 7-93 Long-Range Surveillance Unit Operations/Appendix D.

11. A different look at neural information processing and memory formation. The link between axons, memories and standing waves. April 2018. Jerry Stuger.

12. Group-standing of whistler mode waves near the Moon Y. Tsugawa et al. March 2014.

Chapter 3. Utilizing the Full Moon to Develop One's Intuition

This next chapter shall now concern specific moon phases regarding precognition. Our ARV sessions over the years have found a remarkable enhancement in the accuracy of our ARV sessions during full moons and the first quarter moon (half-moon). There may be an overall cyclic connection with first quarter moons enhancing ARV sessions during solar maximum (and the season from winter to spring) and full moons enhancing ARV sessions during solar minimum (and the seasons from summer to fall), although further research is necessary to confirm this theory.

Edward Leedskalin and Sweet 16

Many people have tried to figure out how Edward Leedskalin was able to move such large stones weighing many tons without using a crane to build his castle in Florida. He gives his readers a clue when he mentions the term "sweet 16" multiple times in his mini-books. Sweet 16 refers to the number of days after the new moon, which gives us the period just after a full moon. Hence we have had remarkable success remote viewing the Dow Jones and FOREX during the full moon. Hence the link between gravity and time. This is just a theory of my own and further research needs to confirm this hypothesis. The other

16 Days 15 Hours old
Waning Gibbous

moon phase that we have had success in remote viewing the financial markets is the first quarter moon. It is interesting to note that the first quarter moon, also called the half moon, is the only moon phase that emits strong polarized light [1] and polarized light exhibits quantum effects [2]. Hence it may be that the stronger polarized light allows access to other dimensions.

Polarized light can also occur from what's known as the 'Brewster's Angle'. This occurs when light reflected from a flat surface of a dielectric (or insulating) medium has partial polarization. This allows the electric vectors of the reflected light to vibrate in a plane parallel to the material's surface. Examples include sheet plastics, undisturbed water, glass and the asphalt of highways.

Further   **Reading**
Coronal Polarization. N.E. Raouafi. Johns Hopkins University Applied Physics Laboratory, Laurel, MD, USA 2010.

Using Polarized Light can Enhance the Clarity of Messages
A reserach study found that the clarity of messages can be enhanced by strengthening its polarity. This is accomplished by spinning the light waves. Fiber optic cables today use light to send messages and the fiber optic lines that are bent or compressed change the polarization of the light, distorting its signal as it travels from point A to point B. This light is also susceptible to sudden weather changes. This is because the light in today's fiber optic cables is un-polarized which becomes easily garbled due to stresses and vibrations placed upon the fiber-optic lines. A search for polarizing the light was undertaken by two physicists who demonstrated a new technique that allows the sending of coded messages using polarization. It uses a fiber ring laser that twirls the light into erratic, yet predictable ways. The key here is that the

polarized light causes much less distortion, making the signal clearer and able to travel much further. In the future this may lead to the ability to speed the rate at which the information is sent. The new technology utilizes a laser which sends streams of fluctuating light down fiber optic lines which branch from the ring [3] [4]. Dr. Ronald L. Mallett also uses spinning light in his time travel research project.

Quote from Dr. Ronald L. Mallett [5]

"My research considers both the weak and strong gravitational fields produced by a single continuously circulating unidirectional beam of light. In the weak gravitational field of an unidirectional ring laser, it is predicted that a spinning neutral particle, when placed in the ring, is dragged around by the resulting gravitational field."

Dr. Mallett also states that at sufficient energies, the circulating laser may produce frame-dragging, as well as closed time like curves (CTC), allowing one to time travel into the past [6], or perhaps even send / receive messages.

Perhaps the moon's phase is affecting brainwaves in such a way it is enhancing alpha brainwaves, which are the prime brainwaves found to occur during PSI or ESP exercises [7] [8].

Further Reading

Spin-Exchange-Relaxation-Free Magnetometry Using Elliptically-Polarized Light. V. Shah and M. V. Romalis. Princeton University. March 2009.

A Summary of the Solar Institute's ARV Sessions

Below is a table of the past few ARV sessions conducted at the Solar Institute regarding our Associative Remote Viewing of the Dow Jones and FOREX Markets.

Reference

www.ez3dbiz.com/pdf_docs/remote_viewing_dow_jones_data_analysis_p df.pdf

Summary of the Data

Out of a total of 23 ARV sessions conducted from February 25th through May 2018 the following data was found

13 Successful

14 Unsuccessful

Hardware Upgrade

A   hardware   upgrade   to   enhance   the   sensitivity   was incorporated into the ARV sessions around mid-may of 2017 increasing our odds of successful ARV sessions slightly.

11 Successful

8 Unsuccessful

A significant boost in accuracy has occurred during our last few sessions during a full moon with low solar wind speeds and low solar activity, especially after a period of very mild solar flare activity.

We at present believe the machine is slightly enhancing the success of our ARV sessions. We do believe that a key factor is from performing HeartMath (heartmath.com) before each ARV session as this seems to greatly enhance the accuracy as it synergizes with the global coherence circuit during these favorable solar weather periods, which we shall cover in greater detail later on. It may be that the ARV device amplifies the present conditions both good and bad, which is why it is key to perform the ARV sessions during the proper moon phase when solar activity is at favorable levels.

Moon Phase Data
Number of Successful ARV sessions at / around first quarter moon (up to 3 days before = 6.5
Number of un-successful ARV sessions at / around first quarter moon (up to 3 days before = 2.5
Number of Successful ARV sessions at / around full moon (up to 3 days before = 2.5
Number of un-successful ARV sessions at / around full moon (up to 3 days before = 1.5
Number of Successful ARV sessions at other moon phases (4th quarter moon and new moons) = 0
Number of Un-successful ARV sessions at other moon phases (4th quarter moon and new moons) = 10

Polar Cap Index
Number of Successful ARV Sessions with increasing / disturbed Polar Cap Index – 1
Number of Successful ARV Sessions with decreasing / quiet Polar Cap Index – 8
Number of un-Successful ARV Sessions with increasing / disturbed Polar Cap Index – 10
Number of un-Successful ARV Sessions with decreasing / quiet Polar Cap Index – 4

Moon Phase Summary

ARV sessions are extremely accurate during the moon's first quarter as well as the full moon when favorable solar conditions are present. This may be a seasonal effect that occurs with the first quarter moon being beneficial for ARV from spring into winter with the full moon delivering favorable results from summer into fall. Further research will confirm this. Extreme solar flare activity that results in a disturbed magnetometer has been found to be detrimental to RetroPK experiments as shown in the Solar-Periodic Full Moon Effect In The Retropsychokinesis Project.

Cosmic Ray Data

Number of Successful ARV Sessions with increasing Cosmic Rays – 10
Number of Successful ARV Sessions with decreasing Cosmic Rays - 0
Number of un-Successful ARV Sessions with increasing Cosmic Rays – 9
Number of un-Successful ARV Sessions with decreasing Cosmic Rays - 4

Our best ARV sessions took place during full moons and first quarter moons and especially so when the sun's solar wind speed was at mild to favourable speeds. This is very exciting because as we shall show later on in this book, solar wind speed plays an important part in health. What is most interesting is higher solar wind speeds have been found to be responsible for hallucinations due to the solar wind influencing pineal hormone and melatonin levels [8b].

This suggests that favourable solar wind speeds create a

more balanced state of mind and perhaps even creating more of a balanced collective unconscious, once again suggesting the effect of a collective unconscious.

It is interesting to note that the most powerful cosmic rays originate from the center of our Milky Way Galaxy (Observation of a large-scale anisotropy in the arrival directions of cosmic rays. A. Aab et al. Sept 2017). The center of the Milky Way is close to the constellation of Sagittarius. The sun transits this sign between approximately November 23 and December 21. The Milky Way itself was practically worshiped by the ancient Maya. It was called the World Tree and was represented by a majestic flowering tree called the Ceiba. It was also called the Wakah Chan, meaning "Six" or "Erect" and "Four", "Serpent" or "Sky". The World Tree was said to be fully grown when Sagittarius was far over the horizon.

The Moon as a PSI Amplifier

Our hypothesis is that the moon acts as an amplifier, not only for emotions, but for the effects of solar conditions that occur when a full or first quarter moon is present. For example if a full moon is occurring and there is major solar flare activity, even if solar wind speeds are mild (between 350 and 400) the accuracy of the ARV session is greatly diminished. On the other hand if a full moon is present and solar activity is mild or non-existent, and solar wind speeds are favourable (350) the accuracy of the ARV session is greatly enhanced. This makes sense because as will be shown later on in this book solar flare activity negatively impacts the body's physiological systems. Studies have found that if the sun's electron flux is low, that it results in an increase in cosmic rays as well as lower atmospheric temperatures [8c]. This would mean that enhanced cosmic rays increase precognition.   We shall explore this connection in greater   detail later on, but first does   the   moon   affect   circulation?

Chakraborty and Ghos theorized that the moon's gravitational pull affects the body's cardiovascular functioning. This causes a person to work slightly harder during full moons causing altered cardiovascular activity [9]. Also some people who are magneto sensitive show elevated blood pressure during increased geomagnetic activity [10] [11].

Wet Cupping Reduces the Severity of Migraines during Full Moons

A research study found that the treating of migraine using blood-letting with wet cupping therapy was significantly better when performed in the second half of the month (full moon) [12]. Wet cupping therapy is believed to remove toxins and harmful substances from the body which promotes healing. Could it be that at full moons a type of detox occurs in the body? It would be interesting to do a study looking at metal chelation therapy and the moon's phase and if more metals are chelated out of the body during full moons. Now let's explore the data to see if cosmic rays are enhancing intuition.

Cosmic Rays and Psychic Ability

Research conducted by Spootswoode [13] found that once per day, a peak in psychic power would take place. The time was 13:30LST. The image is shown on the following page and is courtesy of Apparent Association Between Effect Size In Free Response Anomalous Cognition Experiments And Local

An Introduction to Remote Viewing the FOREX.
Schumann Resonance Coherence Secrets.

Sidereal Time. S. James P. Spottiswoode Cognitive Sciences
Laboratory, Palo Alto, CA 94301.

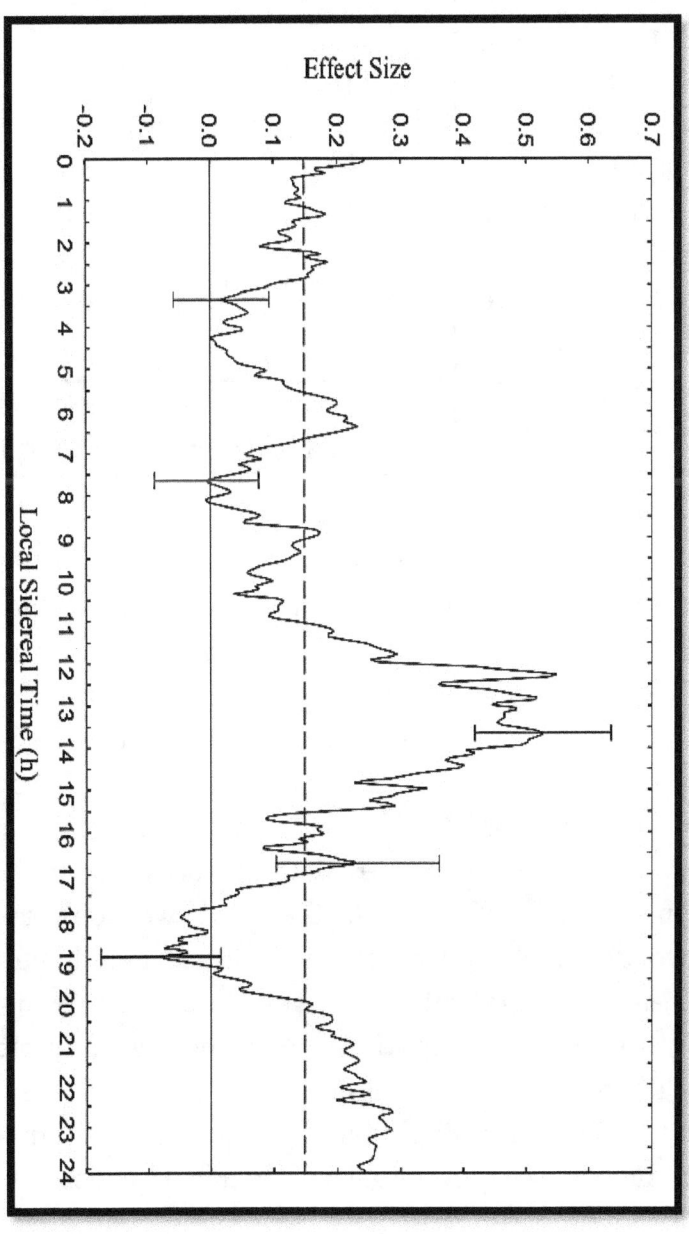

Now if we look at a chart of cosmic rays during 13:30LST, we see a peak taking place. Following image courtesy of Observation of Anisotropy of Cosmic Rays with Solar Time Using the Multidirectional Muon Telescope of GRAPES-3 Shower Array. H. Kojima.

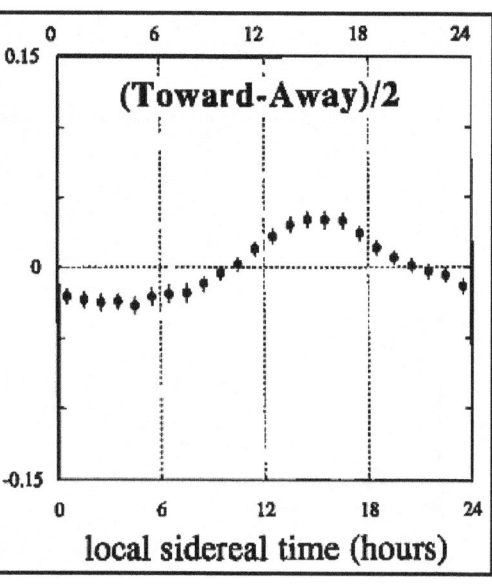

Swinson (1969) found that cosmic ray intensity and its variation reached a maximum peak at 18:00LST and on other days at 6:00LST. The reason for these two peaks is due to a change in polarity (Toward and Away) of earth's interplanetary magnetic field which is based on the flow of cosmic rays that flow perpendicular to the ecliptic plane. The study shows that the there is a distinct peak of the toward flow of cosmic rays at 13:30LST and a low at the away flow of polarity. The paper also states that the peaks have a daily variation at 18:00LST and 6:00LST. If we look at the Spootswoode graph shown earlier, we see a ESP peak at 6:00LST and that the daily low takes place at approximately 18:00LST. These same variations in cosmic rays occurring at 13:30LST has also been observed in other independent studies [14]. The following image is courtesy of Enhanced sidereal diurnal variation of galactic cosmic rays observed by

the two-hemisphere network of surface level muon
telescopes. K. Munakat.

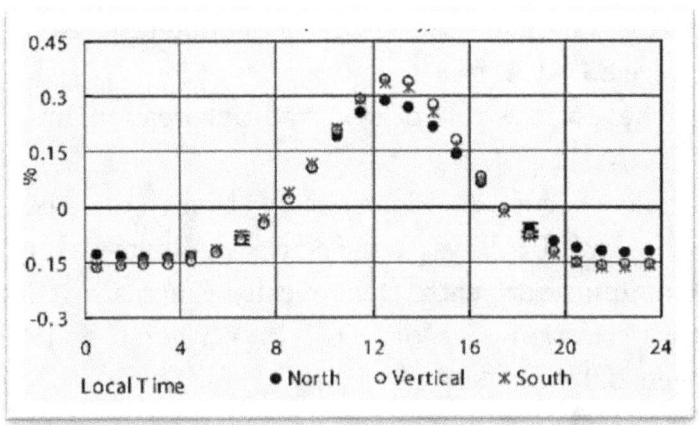

Summary

An increase in cosmic rays may be enhancing the success of
remote viewing sessions, perhaps due to an effect on the
human nervous system. It is interesting to note that a study
looking at cosmic rays and their effects upon the human
nervous system looked at the effect of cosmic rays in the
macula (eyes), thalamus and hippocampus regions of the
brain [16].

As far as our equipment goes, our ARV Amplifier does
seem to be enhancing the results. This combined with
performing ARV sessions during a full moon when solar wind
speeds are favourable and the magnetometer is quiet should
continue to improve the accuracy of our ARV sessions in the
foreseeable future.

References. Chapter 3.

1. Influence of Illumination and Polarized Moonlight on Light-Trap Catch of Caddisflies (Trichoptera)L. Nowinszky. et al. Page 83.

2. Searches for the role of spin and polarization in gravity. Wei-Tou Ni. Page 4.

3. Article: Focus: A Clear Signal from Polarized Light. February 13, 2002· Phys. Rev. Focus 9, 9. Oliver Baker.

4. Communication with Dynamically Fluctuating States of Light Polarization. Gregory D. Van Wiggeren and Rajarshi Roy. Feb 2002.

5. Ronald Mallett. Wikipedia.

6. The Gravitational Field of a Circulating Light Beam. Ronald L. Mallett. Department of Physics, 2152 Hillside Road, University of Connecticut, Storrs, Connecticut 06269 .April 27, 2003.

7. Extrasensory Perception and the Brain Hemispheres: Where Does the Issue Stand Now?Bryan J. Williams. Neuro Quantology | September 2012 | Volume 10 | Issue 3 | Page 350-373 Williams BJ., Extrasensory perception and the brain hemispheres.

8. Psychic Phenomena and the Brain: Exploring the Neuropsychology of Psi' – Psi and the Brain. Williams, B. J. July 7, 2016.

8b. The solar wind and hallucinations--a possible relation due to magnetic disturbances. Randall W1, Randall S. 1991.

8c. Cosmic Influence on the Sun-Earth Environment. Saumitra Mukherjee. Dec 2008.

9. Effects of different phases of the lunar month on humans. Ujjwal Chakraborty. Department of Human Physiology with

Community Health, Vidyasagar University, Paschim Medinipur, West Bengal 721102, India Department of Physiology, New Horizon Dental College and Research Institute, Sakri, Bilaspur, Chattishgarh, India

10. Individual responses of arterial pressure to geomagnetic activity in practically healthy subjects. Zenchenko TA, Dimitrova S, Stoilova I, Breus TK. Klin Med (Mosk). 2009 87(4):18-24.

11. Changing Efficacy of Wet Cupping Therapy in Migraine with Lunar Phase: A Self-Controlled Interventional StudyAli Ramazan Benli and Didem Sunay. Dec 2017.

12. Changing Efficacy of Wet Cupping Therapy in Migraine with Lunar Phase: A Self-Controlled Interventional Study. Benli AR, et al. Med Sci Monit. 2017.

13. Spottiswoode, S. J. P. (1997a). Apparent association between effect size in free response anomalous cognition experiments and local sidereal time. Journal of Scientific Exploration 11(2), Summer. pp. 109 – 122.

14. Sidereal Anisotropy of Galactic Cosmic-Ray Intensity Observed with the Tibet Air Shower Array. K. Kawata for the Tibet Air Shower Array. M. Amenomori et al. 2005.

15. Enhanced sidereal diurnal variation of galactic cosmic rays observed by the two-hemisphere network of surface level muon telescopes. K. Munakat. Feb 1999.

16. Cosmic ray hit frequencies in critical sites in the central nervous system. S.B. Curtis et al. 1998.

17. www.HeartMath.org. Interconnectedness. October 6, 2010. GCI Commentaries. Science Study. www.heartmath.org/gci-commentaries/interconnectedness.

18. The Solar Cycle. David H. Hathaway. Sept 2015.

19. The 10.7 cm solar radio flux (F10.7) K. F. Tapping. March 2013.

Chapter 4. ESP Organs of the body.

Schmeidler (2008) theorizes that the body's lymph nodes and bone marrow which are connected to the body's nerve endings may be the body's ESP organs. During childhood these are more sensitive and active, however age and/or improper diet end up reducing their sensitivity. It is interesting to note that I outline in my book Improve your Remote Viewing Accuracy Techniques using Quantum Microtubules that pineapple, which was suggested by master psychic Doreen Virtue, is a great food to take to enhance psychic ability. Pineapple happens to contain a large amount of manganese (Riordan RD et al. Dec 2004). It just so happens that manganese gathers in the body's lymph nodes. (Synchrotron-based XRF mapping and μ-FTIR microscopy enable to look into the fate and effects of tattoo pigments in human skin. Ines Schreiver. et al. Sept 2017). This is an especially disturbing feature for people who get a lot of tattoos as the ink may sometimes contain manganese, as well as carbon black, which can gather in their lymph nodes. This is because the lymph nodes are very good at holding onto small particles such as dyes or manganese.

Lymph nodes are a part of the body's immune system and are found in the groin, armpits and neck. They act as filters for foreign bodies, keeping away pathogens and cancer cells. Also the material manganese is used heavily in the Remote Viewing Amplifier which is used during Associative Remote

Viewing Sessions.

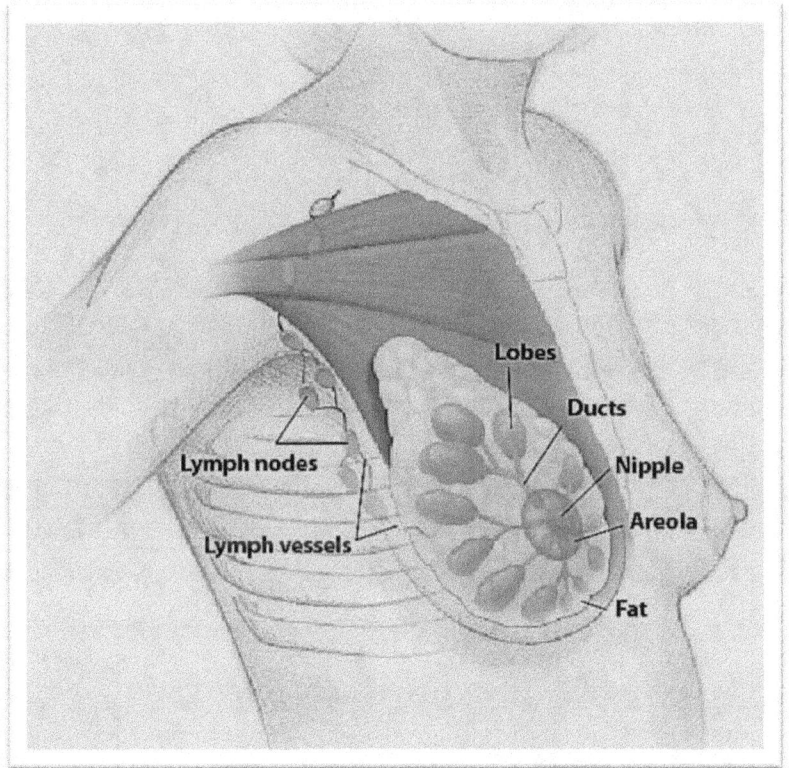

Castor Oil and Lymph Flow

Castor oil was recommended by one of the world's greatest psychics Edgar Cayce. Castor oil helps improve lymphatic flow by enhancing circulation of lymphatic fluids. Apply castor oil to the lower abdomen for best results. Clove is another substance that contains a large amount of manganese (Haizhou Liu. et al. Aug 2014).

It is also interesting to note here that a study involving rats found that when they were fed an oral dose of Limonene, that it reached maximum levels in their lymph glands and lungs after three hours (Distribution and immune responses resulting from oral administration of D-limonene in rats. Hamada M et al. April 2002).

Chapter 5. Solar Weather and Its Effects upon Earth and the Moon.

Earth's Magnetosphere and ESP

During a cycle of solar activity, which lasts an average of 11 days, earth's magnetosphere expands and contracts like a rubber balloon stretching due to the activity of the sun. This creates its own form of pressure. A typical cycle of solar activity includes a high and a low, with the low representing the ending stages of the flare cycle. It is a period where solar wind speeds begin to die down and approach favourable levels. It is during stronger solar activity that earth's magnetosphere stretches into a teardrop shape, which envelopes the sun's solar wind. This teardrop extends far out into space, more than half a million kilometres. Just before a full moon, the moon becomes enveloped in earth's magnetotail (Lichtenstein & Schuert. 1976). The moon can spend up to 3 days in earth's magnetotail, and if a lunar eclipse is present, the moon moves even deeper into earth's magnetotail, close to regions of the plasma sheet. When the moon is not full, it is at the outer ring of earth's magnetotail.

During an average lunar cycle, the moon enters earth's magnetotail 3 days before it turns full. To complete its journey through the magnetotail, the trip takes about 6 days. It may be that the human nervous system becomes more attuned to future phenomena as the moon enters earth's magnetotail. Perhaps while enveloped in the magnetotail a type of shielding mechanism goes into effect. It is interesting to note that in the published paper titled: Solar Periodic full moon effect in the fourmilab retropsychokensis project by Eckhard Etzold, that it showed that during stronger solar activity around the time of the full moon, RetroPK activity was greatly reduced. This means that enhanced solar radiation may be causing some kind of interference.

Manganese and gingko are the two substances we use in our ARV sessions. The food TEFF, which is high in Manganese is taken 5 hours before the ARV session for dinner and Gingko about 5 minutes before the ARV session begins. Mung beans and chickpeas are also high in manganese as well as polyphenols (Polyphenol-Rich Lentils and Their Health Promoting Effects. Kumar Ganesan and Baojun Xu. Nov 2017). It just so happens that polyphenols protect DNA against radiation (Protective action of plant polyphenols on radiation-induced chromatid breaks in cultured human cells. Parshad R. et al. Oct 1998). As a matter of fact Gingko is such a powerful protector against negative types of radiation that a 390 year old Gingko tree, known as the Yamaki Pine, was the only tree left standing after the atomic bombing of Hiroshima (The 390-Year-Old Tree That Survived the Bombing of Hiroshima. Katie Nodjimbadem. www.smithsonian.com August 4, 2015). If this is true, it means when solar activity is stronger during a full moon, that nature may be exerting a type of cleansing effect by exerting certain types of solar radiation.

Cycles of the Sun's Solar Wind

The sun's solar wind speed has a distinct cycle, much like the 11 year sunspot cycle. Before we explore further, I would like to clearly point out here that after many years of remote viewing the future position of the dow jones (ARV), our most successful ARV sessions would always take place during the winter / spring months. As will be shown, the sun's solar wind speed happens to enter its lowest speed / favourable speed (between 330 and 350) during winter each year (nov to dec) exhibiting a seasonal variation. Hence the sun's solar wind speed consists of a distinct seasonal cycle. Also during solar minimum, which occurs approximately every 11 years on average, solar wind speeds are usually lower (Cliver & Ling 2011 de Toma 2011). As the following image shows, the solar wind speed goes into a decline / lower speed approaching solar minimum. The following image is courtesy of the paper titled: Near-earth solar wind flows and related

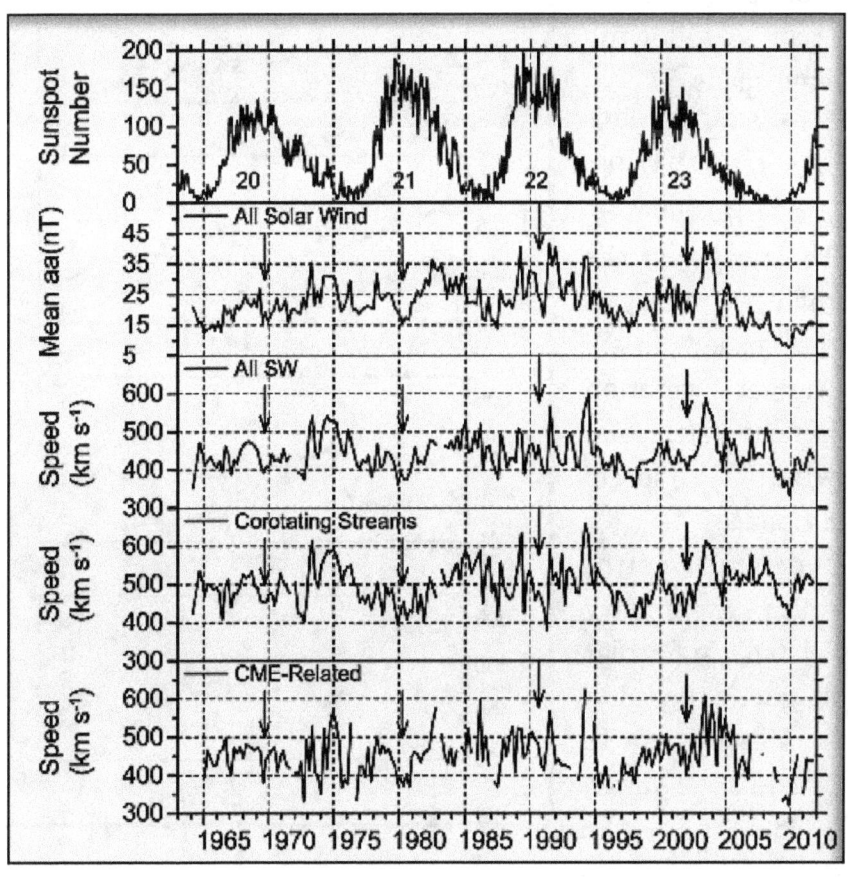

geomagnetic activity during more than four solar cycles (1963–2011). Ian G. Richardson and Hilary V. Cane. Oct 2003.

This trend has been shown to occur in another independent study. As the previous image showed, solar wind speed during sunspot minimum is lower on average. The next image shown below right displays 27-day averages of solar wind speeds for two years during the approximate last four sunspot minimums which took place during the following time-spans - (a) 1965-66, (b) 1975-76, (c) 1987-88, and (d) 1994-95. The image on the right is courtesy of the paper titled: Annual variation in near-Earth solar wind speed' Evidence for persistent north-south asymmetry related to solar magnetic polarity. B. Zieger and K. Mursula. March 1998.

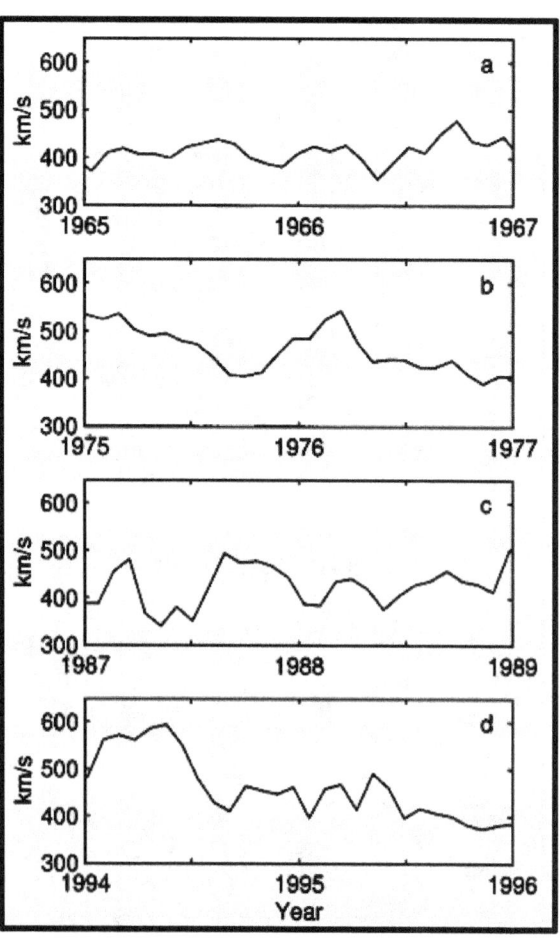

The solar wind speed also exhibits a grand cycle that lasts approximately 87 years. The last minimum took place in the first decade of this century (Geomagnetic and solar wind cycles, 1900–1975. Aug 1982).

The 2 Main Speeds of the Sun's Solar Wind

The sun's solar wind emerges from magnetic fields in its heliospheric current sheets that separate both hemispheres of the sun. These have opposite polarities from one another. The solar wind has two speeds that can emerge from any hemisphere of the sun. These speeds are fast and slow. The fast solar wind speed (above 500) comes from large coronal holes. These holes occur most often during a declining sunspot cycle headed towards sunspot minimum. The fast solar wind speed can also result from coronal mass ejections and solar flares during solar maximum [1].

Cycles of Solar Wind Speeds

As the sunspot cycle begins to decline, headed towards sunspot minimum, the coronal holes at the polar regions of the sun extend outwards towards the sun's equator. If it happens to be the season of fall or spring during this time, the solar wind speed will reach maximum speeds. Research by Mursula and Zieger (1998) discovered that during positive polarity sunspot minimum, the sun's solar wind speed was faster around the spring equinox, compared to the solar wind speed during fall. During negative polarity around sunspot minimum, the solar wind speed will become faster around the fall equinox compared to the solar wind speed during spring. This occurs because an accumulation of momentum takes place in earth's atmosphere during winter, which becomes released during spring (Krymskij (1993). The following image shows a graph showing solar wind speeds during these cycles with positive polarities being a sold line and negative polarities being the broken line. Hence, solar wind speed can reach higher speeds even during solar minimum, and is more likely to do so at the equinoxes [2]. It is interesting to note an overall slower solar wind speed appears to take place during winter time (December) each year suggesting a seasonal variation, although further studies are necessary to see if this variation is constant [3]. The following data is from OMNI between 1965 and 1999. This may explain why a maximum of solar wind speed activity took place during the 1952 equinoxes in the declining phase of solar cycle 18.

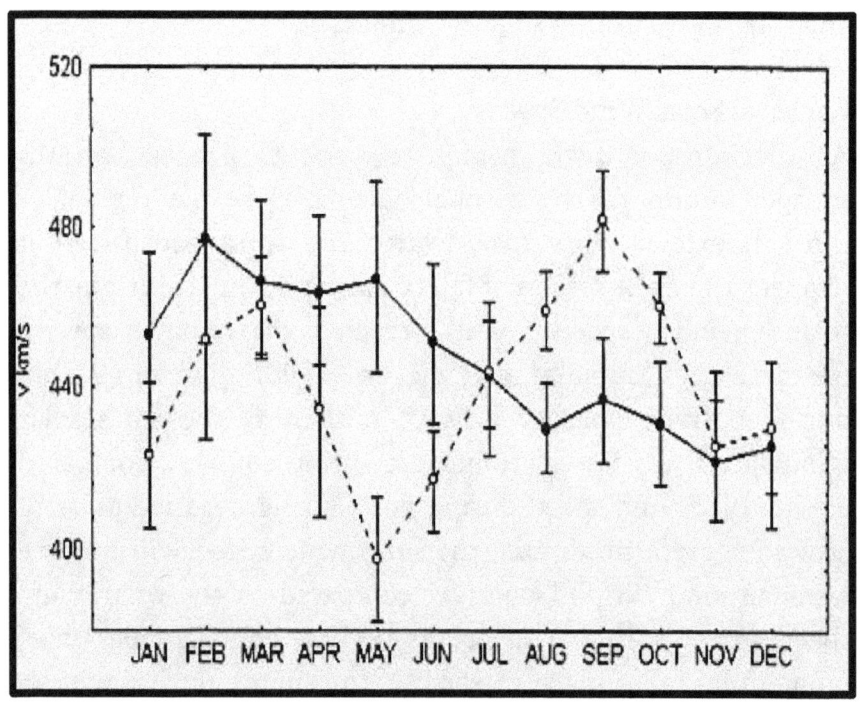

An Introduction to Remote Viewing the FOREX.
Schumann Resonance Coherence Secrets.

The above image is courtesy of: Space Science: New Research - Nick S. Maravell. Page 60.

The next image is the sun's solar wind speed from a separate study conducted from 1985 to 2010 titled: Heliolatitude and Time Variations of Solar Wind Structure from in situ Measurements and Interplanetary Scintillation Observations. J.M. Sokól et al. April 2012. Page 170

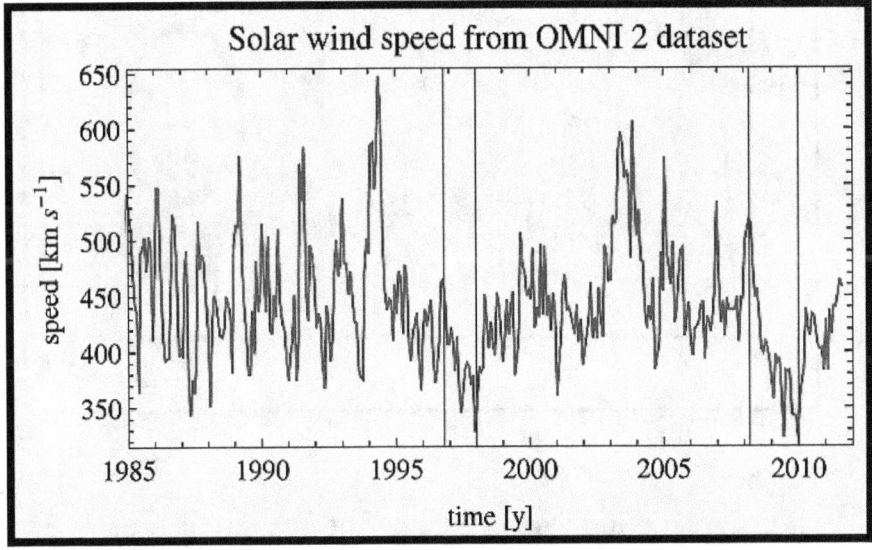

What's interesting to note is that if we look at sunspot cycle number 23 in the following image, we see that it is of a lower height than the previous sunspot cycle 22. Hence this may be what created the extra low solar wind speeds from 1997 to

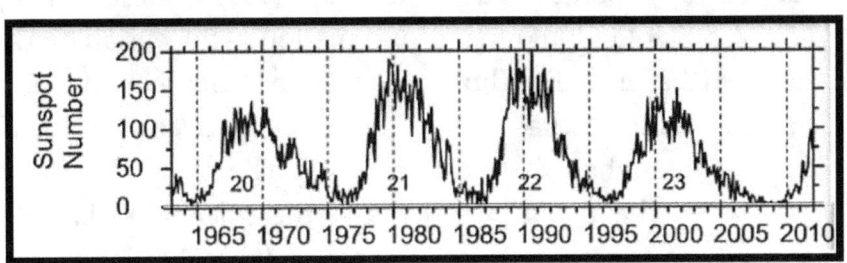

1998. The years 1997 to 1998 are the years that the general public became interested in remote viewing, most of which was broadcast via Coast to Coast AM by Art Bell.

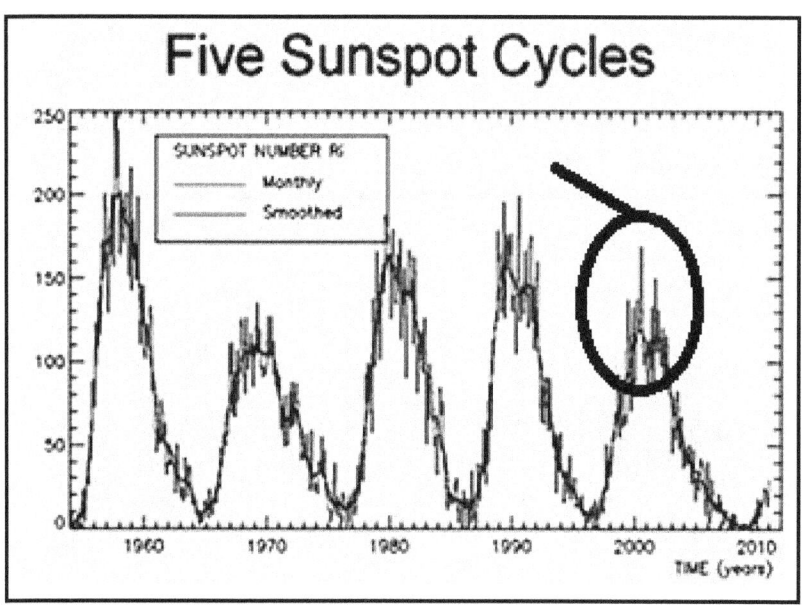

Further   **Reading**
Global solar wind variations over the last four centuries. M. J. Owens. Jan 2017.

The Solar Wind, Full Moons and RetroPK

The solar wind speed and its effects on cognition have been studied by Spottiswoode and May (1997). Research by Eckhard Etzold (2005) found that the following variables: the sun's F10.7cm radio flux, a deviation from ecliptic, and sunspot activity had the greatest impact on RetroKP experiments when the moon was full. The study also found that when the moon was not full, **the solar wind had the greatest impact on RetroPK performance**, followed by geomagnetic activity levels with minor effects attributed to the sun's 10.7cm radio flux (F10.7) [4].

Variables from best to last - Full Moon (solar minimum)

1 - Solar Wind Speed (low)
2 - Geomagnetic Activity (low)

Variables from best to last - Non Full Moon (solar maximum)

1 - Solar wind speed
2 - Geomagnetic activity levels
3 - The sun's 10.7cm radio flux (F10.7)

The 2 Main ARV Cycles

ARV Cycle #1 - Solar Maximum / increasing sunspots + First Quarter Moon = ARV sessions work best (RetroPK sessions may also show this effect, although further research is needed).

ARV Cycle #2 - Solar Minimum / declining sunspots + Full Moon = ARV / RetroPK Sessions work best, unless solar flare activity is present. Above average geomagnetic activity can exhibit negative effects during this time. Favourable / low solar wind speeds (330 to 350) may be more beneficial during this time.

What does Deviation from the Elliptic Mean?

Deviation of the ecliptic is the inclination of a planet's equator with respect to the ecliptic, or its rotation axis to a perpendicular to the ecliptic. For example earth's is

approximately 23.4° and is decreasing at the rate of 0.013 degrees (47 arc seconds) every hundred years [5] [6]. Speaking from personal experience, I have achieved the very best ARV sessions when the solar wind speed happened to be at 330 and geomagnetic activity was low to quiet with a major boost when the sun's 10.7cm radio flux was increasing during solar minimums. This also makes sense, because the majority of our ARV sessions took place when the moon was not full and during ARV cycle #2. It would also explain the failure of the April 3rd, 2018 ARV session which took place around a full moon when a solar flare occurred (an increase in the sun's 10.7cm solar flux and a similar effect to a solar maximum). Hence the effects of the strong solar activity shielded the beneficial "charge" from the moon. Shown below is the April 3rd, 2018 major solar flare which also resulted in an increased

/

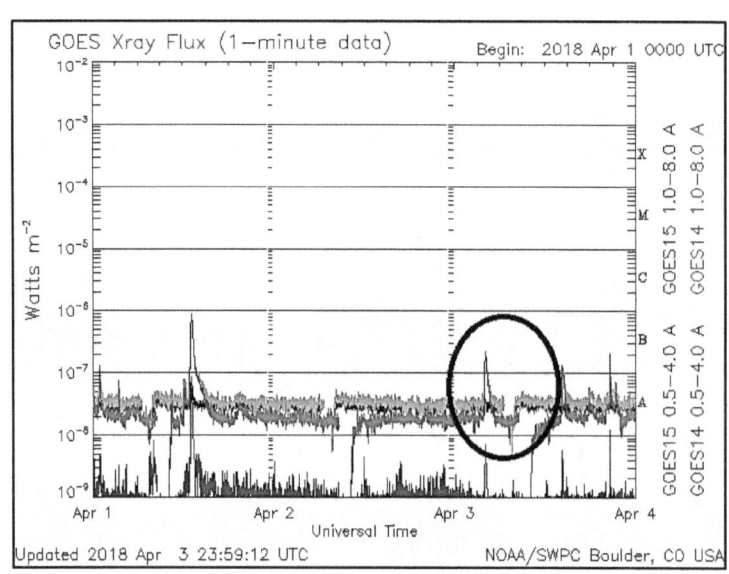

disturbed solar flux.

Solar Flux Reference http://legacy-
www.swpc.noaa.gov/ftpdir/warehouse/2018/2018_plots/xray/
20180403_xray.gif

Shown below is sunspot activity

```
#              Quarterly Daily Solar Data
#
#                    Sunspot        Stanford GOES15
#              Radio  SESC   Area              Solar  X-Ray  ------ Flares -----
#              Flux  Sunspot 10E-6    New      Mean   Bkgd    X-Ray      Optical
#  Date       10.7cm Number Hemis. Regions    Field  Flux   C M X S    1 2 3
#-------------------------------------------------------------------------------
2018 04 01      69      0      0       0       -999   A0.0   0 0 0 1    0 0 0
2018 04 02      68      0      0       0       -999   A0.0   0 0 0 0    0 0 0
2018 04 03      68      0      0       0       -999   A0.0   0 0 0 3    0 0 0
2018 04 04      69      0      0       0       -999   A0.0   0 0 0 0    0 0 0
2018 04 05      66      0      0       0       -999   A0.0   0 0 0 0    0 0 0
2018 04 06      67      0      0       0       -999   A0.0   0 0 0 0    0 0 0
```

Ref   http://legacy-
www.swpc.noaa.gov/ftpdir/indices/old_indices/2018Q2_DSD.
txt

Shown below is earth's KP activity measured in
Fredericksburg K-Indices

```
#
#           Middle Latitude      High Latitude        Estimated
#          - Fredericksburg -    ---- College ----   --- Planetary ---
# Date      A    K-indices        A   K-indices       A   K-indices
2018 04 01  4  1 1 2 1 2 1 1 1    3  1 0 2 2 2 0 1 0   5  2 1 1 2 2 1 1 1
2018 04 02  4  1 1 1 0 2 2 1 1    2  1 1 0 2 1 1 0 0   5  2 1 1 1 2 2 1 1
2018 04 03  3  1 2 2 1 0 1 1 0    2  1 1 2 0 0 0 0 0   4  2 2 2 0 0 0 0 0
```

Ref   -   http://legacy-
www.swpc.noaa.gov/ftpdir/indices/old_indices/2018Q2_DGD
.txt

Below is the solar wind speed on April 3rd, 2018.

```
#    1-minute averaged Real-time Bulk Parameters of the Solar
#
#                   Modified Seconds    -------------   Solar Wind
# UT Date   Time   Julian   of the              Proton   Bulk
# YR MO DA  HHMM    Day      Day      S   Density   Speed
#------------------------------------------------------------
2017 04 03  1902   57846    68520    0      3.0     412.4
2017 04 03  1903   57846    68580    0      3.2     413.5
2017 04 03  1904   57846    68640    0      3.0     411.9
2017 04 03  1905   57846    68700    0      2.6     412.2
2017 04 03  1906   57846    68760    0      2.9     410.9
2017 04 03  1907   57846    68820    0      2.7     411.6
```

Ref   http://legacy-
www.swpc.noaa.gov/ftpdir/lists/ace/20170403_ace_swepam_1
m.txt

The ARV session conducted on **Tuesday Evening May 29th, 2018** was a success and the solar weather details are shown on the following pages.

Date	Radio Flux 10.7cm	SESC Sunspot Number	Sunspot Area 10E-6 Hemis.	New Regions	Stanford Solar Mean Field	GOES15 X-Ray Bkgd Flux	Flare X-Ray C	M	X	S
2018 05 26	73	26	70	0	-999	A4.5	0	0	0	0
2018 05 27	75	27	50	0	-999	A4.1	0	0	0	1
2018 05 28	77	20	80	0	-999	A4.8	1	0	0	7
2018 05 29	75	22	70	0	-999	A4.1	0	0	0	2
2018 05 30	75	18	30	0	-999	A3.7	0	0	0	0
2018 05 31	77	21	50	0	-999	A6.0	0	0	0	2
2018 06 01	75	22	60	0	-999	A6.7	0	0	0	1

Shown above is sunspot activity with an S flare on the 28th
Ref   http://legacy-
www.swpc.noaa.gov/ftpdir/indices/old_indices/2018Q2_DSD.
txt

Shown below is solar flare activity with the disturbed solar flux representing the S flare. (The Sun's Solar Flux)

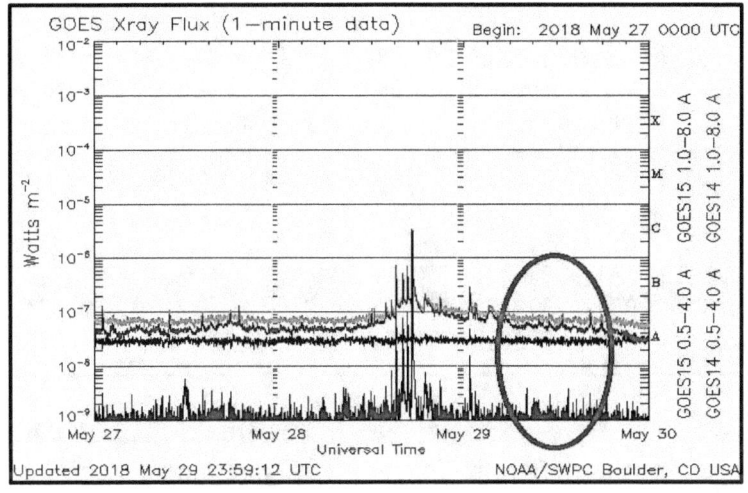

Ref http://legacy-www.swpc.noaa.gov/ftpdir/warehouse/2018/2018_plots/xray/20180529_xray.gif

The following image are the Middle Latitude Fredericksburg K-indices.

Date	Middle Latitude - Fredericksburg -		High Latitude ---- College ----		Estimated --- Planetary ---	
	A	K-indices	A	K-indices	A	K-indices
2018 05 27	5	2 2 1 2 2 1 1 1	5	2 2 0 3 2 0 1 1	4	2 2 1 1 1 1 1 1
2018 05 28	4	1 1 1 2 2 1 1 1	2	0 1 1 1 1 0 1 0	4	1 1 1 1 2 1 1 1
2018 05 29	5	2 2 1 2 2 1 1 1	2	1 2 1 0 0 0 1 0	4	1 2 1 1 1 1 1 1
2018 05 30	5	1 1 1 1 2 2 2 2	2	0 1 1 0 0 0 2 1	4	1 1 1 1 1 1 2 2
2018 05 31	11	2 2 1 2 3 3 3 3	9	2 2 0 3 1 3 3 2	12	2 2 1 2 2 4 4 3

Ref http://legacy-www.swpc.noaa.gov/ftpdir/indices/old_indices/2018Q2_DGD.txt

Shown below is the sun's solar wind speed.

```
#   1-minute averaged Real-time Bulk Parameters of the Solar
#
#                     Modified Seconds   -------------  Solar Win
# UT Date    Time   Julian  of the                Proton    Bulk
# YR MO DA   HHMM    Day     Day     S    Density     Speed
2017 05 29   2035    57902   74100   0        2.1       372.6
2017 05 29   2036    57902   74160   0        1.7       359.9
2017 05 29   2037    57902   74220   1        1.4       350.2
2017 05 29   2038    57902   74280   0        2.4       370.7
2017 05 29   2039    57902   74340   0        2.5       374.2
2017 05 29   2040    57902   74400   0        2.5       373.8
```

Ref   http://legacy-
www.swpc.noaa.gov/ftpdir/lists/ace/20170529_ace_swepam_1
m.txt

Summary

The disturbed solar flux resulting from solar flare activity acts as an amplifier. The May 29th, 2018 ARV session was a success because the enhanced solar flux was associated with favorable solar wind speeds of approximately 350. The April 3rd, 2018 ARV session was a failure because the solar wind speeds were at approximately 410 when the ARV session was conducted, which is out of the favorable 350 range. The influence of the sun's solar wind is the determining factor in the success of RetroPK when a full moon is not present. During full moons the sun's F10.7cm radio flux, and deviation from elliptic are favoured positions for successful RetroPK sessions. When the moon is not full, lower solar wind speeds are favoured for successful RetroPK sessions.

Highlighted solar data courtesy of the online Dow Jones Remote Viewing Project located at: www.ez3dbiz.com/dow_project_research_summary.html

Warehoused data
http://legacy-www.swpc.noaa.gov/ftpdir/warehouse/

http://legacy-www.swpc.noaa.gov/ftpmenu/indices.html

The Solar Radiation Shielding Effect

The reason ARV sessions and RetroPK sessions fail during times major solar flares occur which result in a large disturbance of the sun's solar flux and magnometer during full moons could be attributed to the cosmic ray shielding effect, where stronger solar activity results in a shielding type effect. This shield results in less cosmic rays penetrating earth's atmosphere, shielding the moon's electromagnetic energy. Hence, RetroPK and possibly ARV performance is regulated by a delicate balance of solar activity and geomagnetic activity. It also suggests a DNA link to remote viewing with DNA perhaps acting as an antenna. This is because cosmic rays may be affecting our DNA [7].

Cosmic rays affecting humans have a bad reputation due to their effects on astronauts in outer space. Without earth's atmosphere to shield us from cosmic rays we would undergo severe health problems. It may be that a 'sweet spot' of cosmic ray activity exists due to the shielding of earth's atmosphere. This sweet spot being an increase in cosmic rays, especially as excess solar activity starts fading. Because we have earth's atmosphere as well as enhanced solar activity to shield us from the increased cosmic rays, it may be creating a sweet spot that is beneficial for remote viewing. As we shall show in great detail throughout this book, **enhanced parasympathetic activity** is one of the keys to successful associative remote viewing. This would suggest a link between HeartRate Variability and cosmic rays. A research study has found that **increased cosmic rays** are associated with increased HRV as well as **parasympathetic activity** [8]. Cosmic rays follow a 22 year cycle (Webber and Lockwood, 1988). It is interesting to note that the solar sunspot cycle is approximately 11 years [9] which is half that of the cosmic ray cycle. The following image shows that when sunspots increase, cosmic rays decrease. The following image is courtesy of: Solar Activity, Cosmic Rays, and Earth's Temperature: A millennium-scale comparison. I. G. Usoskin.

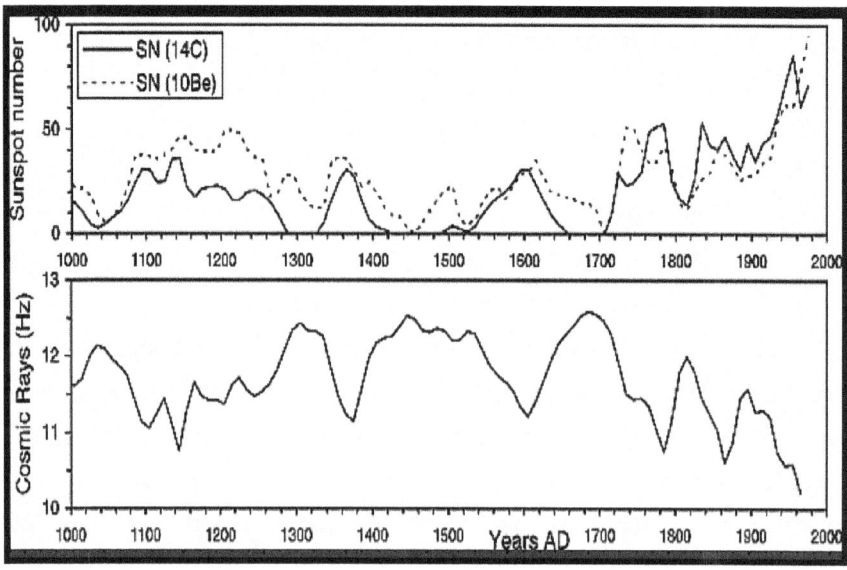

et al. Oct 2005.

Cosmic Rays and Computer Malfunctions

As cosmic rays collide with earth's upper atmosphere, the high-energy neutron particles they produce combine with atomic nuclei. During 2004, researchers from IBM measured the flux of neutrons from cosmic rays and discovered that the error rate in computer memories increased when cosmic rays increased [10].

These types of errors are known as 'soft errors' in which a signal or datum is wrong, but not serious enough to cause long-term permanent damage. In spacecraft these 'upsets' are known as single event upsets and are remedied by cold booting the computer's operating system. Soft errors don't damage a system's hardware only the data that is being processed. These soft errors occur as high energy space particles hit the computer's chip. This causes the radioactive atoms in the computer chip's rate of decay to release alpha particles which flood into the chip. Alpha particles contain positive charges, as well as moving kinetic energy. This causes the particles to enter memory cells causing a change state which causes a difference in values. This reaction is atomic in scale so it does no actual physical damage to the physical structure of the chip (Ziegler and Lanford, 1979).

Hence, if background levels of neutrons explain errors in computer memory it should explain errors in DNA replication as theorized by González. González has based his work on experiments conducted by Richard Lenski and collegues at Michigan State University since 1988. Their team is growing E. coli and monitoring mutations that occur between each generation. He states high-energy neutrons enter the water and collide with the bacterial samples every 125 seconds. The neutrons then transfer their energy into the water molecules which creates a short track of ions. He says that just a single neutron is enough to generate 300 ions over a length of approximately 100 nanometers with about 30 ions at a total distance of 0.1 mm. Hence, bacteria experiencing this ion shower may experience mutations, be destroyed or experience permanent damage, especially in their DNA, which is than inherited by their descendants [11]. This is a fascinating study because it shows that at the nanoscale, bacteria is being affected and influenced by cosmic rays.

The number of cosmic rays is dependent upon altitude which is why computer systems located underground show a reduction in soft errors. However computers on top of mountains, such as those that manage large space telescopes, experience higher rates of soft errors compared to sea level. Soft errors in aircraft occur up to 300 times more than that of sea level upsets. As the density of computer chips increase, Intel expects errors caused by cosmic rays to grow and eventually become a limiting factor in design [12].

Cosmic-ray soft errors show an inverse proportion to sunspots. **Soft errors decrease towards solar maximum and increase towards solar minimum** [12]. Could this mean that environments where soft errors occur be environments favorable to remote viewing? We show throughout this book that enhanced cosmic rays actually increase the success of remote viewing. Further research is necessary to confirm this theory.

During 2004 IBM researchers measured the flux of neutrons from cosmic rays and used the data to predict soft errors in computer memories. Their predictions matched the observed rate of error [13].

10B Boron and Cosmic Rays.

There are 19 different types of boron. Born 10 and 11 are the only two that are the most stable. It also happens to be that naturally occurring boron is 20% 10B with the remainder being 11B. The majority of soft errors are caused by high levels of 10B which occur in the lower layer of older integrated circuits. However when Boron-11 is used at low concentrations (a p-type dopant) the soft error rate is greatly reduced. As a matter of fact integrated circuit manufacturers got rid of borated dielectrics as circuit components decreased in size to 150 nm because of the extremely large rates of soft errors that were occurring. Boron-11 happens to be a by-product of the nuclear processing industry [14]. This may mean that Boron 10 which is freely available may enhance the rate at which cosmic rays enter an environment. Further research is necessary to conform this theory.

Soft Errors are Beneficial in Medical Diagnostics

We have all seen the movie terminator where the time travelling robots fight one another in front of an MRI machine. High energy cancer radiation therapy creates neutrons which become scattered from the walls and equipment in the room causing a thermal neutron flux which is approximately 40×106 higher than normal. This results in very high rates of soft errors [15] [16].

Summary
Because cosmic rays affect computer memory, it may be that when cosmic rays increase, the brain is able to hold and process more information. It would be interesting to do brain studies and observe how fast the brain processes information depending upon the number of cosmic rays and see if more information is processed faster when there are more cosmic rays. What is also interesting is the sun's 10.7 cm solar radio flux is the frequency of 2.8 GHz. 2.8 GHz happens to be the clock speed of some computer processors.

Cosmic Rays and Atmospheric Changes

Types of Cosmic Rays
Like the solar wind speed, there are two different types of cosmic rays [18].

Multiple-muons
Multiple-muons come from primary cosmic rays that have energy higher than the one needed for single muons.

Single-muons
Single muons are cosmic rays of a lower energy than multiple-muons.

Single muons peak during summer [18].

Multiple muons peak during winter [18].

Cosmic ray variations caused by differences in air pressure have been mainstream science knowledge for a long time.

Steinke (1929) and Myssowsky and Tuwim (1926) were the first to study the relationship between cosmic ray time variations and atmospheric pressure changes.

Because less solar activity means more cosmic rays, we have found that our ARV sessions go well when the barometric air pressure has peaked and is starting to decline over the next few days. As the bottom part of the lower air pressure approaches, rain is usually more common. Another time rain is more common is when cosmic rays increase. Cosmic rays also have an effect on cloud formation. A study found that during geomagnetic storms (usually accompanied by stronger solar activity) there were less clouds during September and March (equinoctial months). The study concluded that the phenomenon indicates an influence of cosmic rays and cloud formation [18]. This is interesting because as we covered in an earlier chapter, research by Mursula and Zieger (1998) discovered that during positive polarity sunspot minimum, the sun's solar wind speed was faster around the spring equinox, compared to the solar wind speed during fall.

Seasonal Cycles of Cosmic Rays
Variations in cosmic rays arise from changes in temperature caused by the change of seasons which in turn cause a change in the density of earth's atmosphere.

Varying Changes in Atmospheric Pressure
During winter the atmosphere is cooler, more shallow and dense. Cosmic ray interactions happen closer to the Earth's surface, in a more dense environment. As temperatures increase, earth's atmosphere becomes less dense. This means

the probability for mesons to interact with molecules in earth's atmosphere is greatly reduced. Corresponding increases in meson decay create a larger intensity of muons during summer. Hence, as shown in the graph below, there are more cosmic rays during summer (single muon) with the gradual increase beginning during spring.

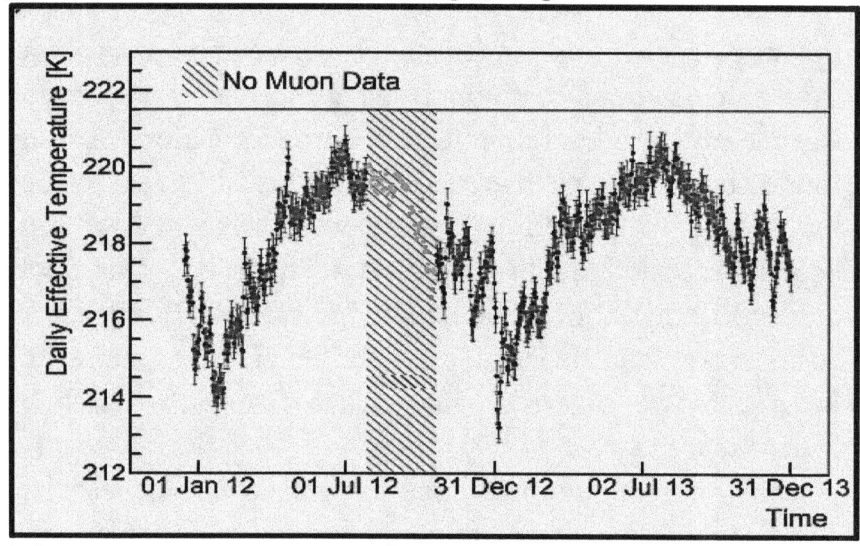

Prior image courtesy of Seasonal Variation of the Underground Cosmic Muon Flux Observed at Daya Bay. The Daya Bay Collaboration. Jan 2018.

Atmospheric Temperature Affects Cosmic Rays

Experiments have now verified that cosmic ray intensity is positively correlated with atmospheric temperature and that the corresponding increase in the decay of cosmic rays yields larger muon intensities during summer months [19]. In another study scientists discovered that **atmospheric pressure affects cosmic rays** [20].

An Introduction to Remote Viewing the FOREX.
Schumann Resonance Coherence Secrets.

Quotes   from   the   study   -

'The sharp increase of cosmic rays may be due to the sudden drop of atmospheric pressure '
Solar flares and magnetic storms suppressed our cosmic ray counts'

The study concluded that atmospheric stability can cause major changes in cosmic ray activity.

Summary
ARV sessions are much more accurate when barometric air pressure has peaked and is starting to decline.   Air Pressure is shown on the top graph and Cosmic Rays are the bottom graph

In a related independent study we can see that as barometric air pressure drops, cosmic rays increase. This is shown in the following   image.

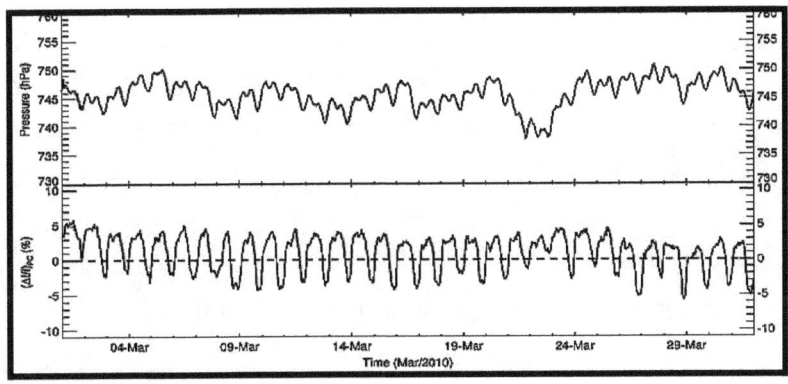

Above image courtesy of Analysis of atmospheric pressure and temperature effects on cosmic ray measurements. R.R. S.

De Mendonça. et al. April 2013.

Solar Cycles and Cosmic Ray Intensity

During Solar Minimum, which is a time solar flare activity is minimal there are more cosmic rays.

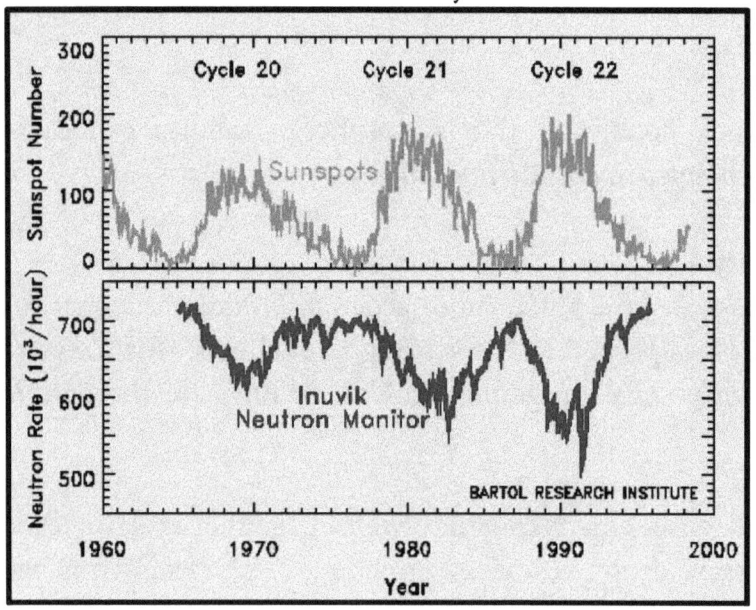

Above image courtesy of The Inuvik Neutron Monitor showing solar cycles 20 through 22 (neutronm.bartol.udel.edu). As of 2018 we are in a cycle where sunspots are in decline, hence we are seeing an increase in cosmic rays.

Long Term Solar Cycles

The 11-year sunspot peaks grow higher and higher for a period of forty years then fade away to complete an 80-or 90-year cycle. This has been confirmed by the German botanist Schnelle.

Reference

The Solar Cycle. David H. Hathaway. Mail Code VP62. NASA
Marshall Space Flight Center. Huntsville, AL 35812, U.S.A.
February 2010.

Cycles   of   Geomagnetic   Storms

The intensity of geomagnetic storms follows the solar cycle
[22]. The below image shows long-term intense geomagnetic
storm events and the solar cycle. Below image courtesy of
Long-term occurrence probabilities of intense geomagnetic
storm events K. Tsubouchi and Y. Omura. April 2007.

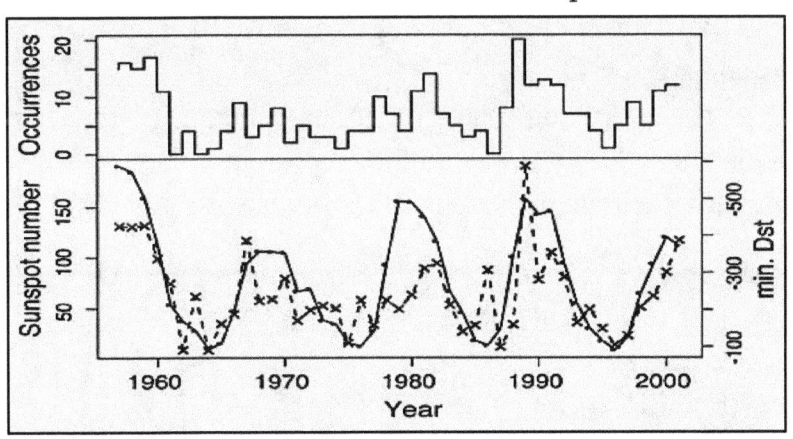

Another study found that 27% of geomagnetic storms
occurred towards solar maximum and 73% of the
geomagnetic storms would take place towards solar
minimum. Also 90% of the strongest geomagnetic storms
occur 2 years before solar maximum or three years after solar
maximum [22].

Geomagnetic activity is twice as strong during winter
compared to summer with the summer geomagnetic storms
lasting about an hour longer on average compared to winter

[23]. Additional large variations have been found to take place in Spring 1994 and Fall 2003, which are both periods of declining solar activity (cycles 22 and 23). Also geomagnetic activity exhibits an approximate 27 day cycle that comes from solar wind structures recurring each solar rotation [25].

Chernosky (1966) observed a long term 22-year cycle geomagnetic storm cycle. Their effects are explained in greater detail by Russell, 1975 and Svalgaard, 1977. It is interesting to note that the sun's 10.7cm solar radio flux has a 10 to 13 year cycle with the sunspot solar activity cycle [25]. This is approximately half of that of the geomagnetic activity cycle. Below is a chart showing the sun's 10.7cm solar radio flux and sunspots [26].

In the following image, aa-index is geomagnetic activity and is the middle graph and Rz (Wolf number) is sunspot number and is the bottom graph.

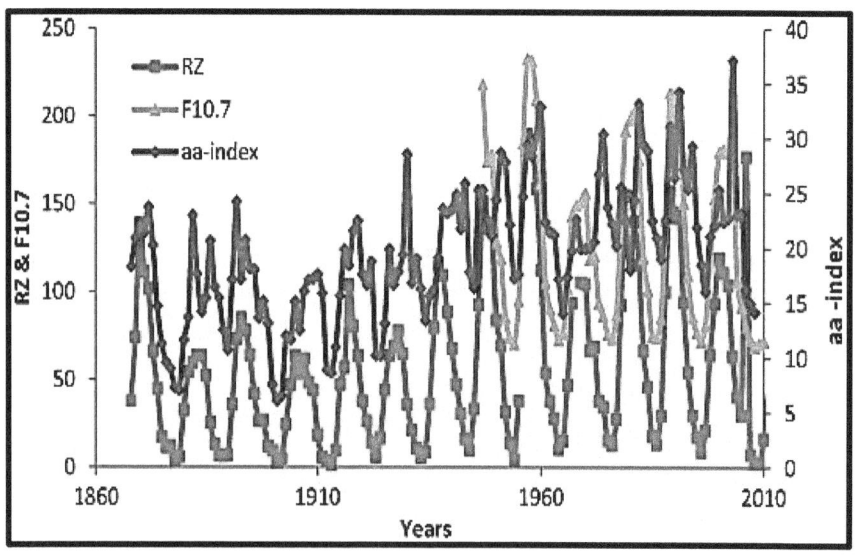

Previous image courtesy of Forecasting the Peak of the Present Solar Activity Cycle 24. Author links open overlay panel. R. H.Hamid and B.A.Marzouk. March 2018.

Further Reading

The KP index and solar wind speed relationship: Insights for improving space weather forecasts. Heather A. Elliott. et al. April 2013.

Cycles of Geomagnetic Activity and Moon Phase

During and on approach to solar minimum (ARV cycle #1) earth's geomagnetic activity (KP) levels are higher on average during new moons and lower on average during full moons [26]. During solar maximum, minimum values of KP activity occur several days before the full moon with maximum KP activity occurring several days after the full moon [27] [28]. So we can see that ARV cycle #1 is the best cycle to practice ARV sessions. This is due to the fact that full moons enhance RetroPK, solar wind speeds are lower during solar minimum and earth's geomagnetic activity is lower around the time of a full moon. Lower geomagnetic activity and lower solar wind speeds all contribute to enhanced success of remote viewing sessions. This is because geomagnetic activity closely follows solar wind speed [29] and geomagnetic activity peaks each year during the equinoxes [30] [31]. Below is a graph showing the peaks in geomagnetic storms at the equinoxes.

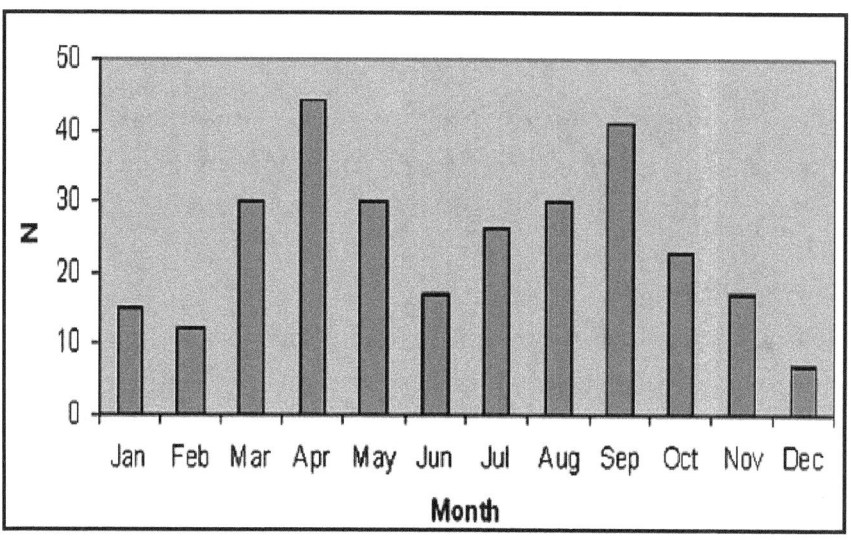

Final   **Summary**

During solar maximum (ARV cycle #2), the full moon may also not be as necessary for successful ARV sessions because these are periods a stronger 10.7 cm solar radio flux occurs more often. Also large variations in solar wind speed take place around solar maximum (large numbers of sunspots).

The 2 Main ARV Cycles

ARV Cycle #1 - Solar Maximum / increasing sunspots + First Quarter Moon = ARV sessions work best (RetroPK sessions may also show this effect, although further research is needed).

ARV Cycle #2 - Solar Minimum / declining sunspots + Full Moon = ARV / RetroPK Sessions work best, unless solar flare activity is present. Above average geomagnetic activity can exhibit negative effects during this time. Favourable / low solar wind speeds (330 to 350) may be more beneficial during this time.

Locating the Most Favourable Solar Weather Conditions for an ARV Session

While the details and data are out of the scope of this book, I explain them in more detail in my book Improve your Remote Viewing Accuracy Techniques using Quantum Microtubules, I will give a general idea on how to find favourable solar weather conditions.

One simple clue to check solar activity is to look at the most recent activity of cosmic rays. If it shows an increase than it is very likely that solar activity is lower which is advantageous during ARV Cycle #2 as long as solar wind speeds are not extremely high, Middle Latitude Fredericksburg K-indices are between 4 and 11 and the magnometer is not terribly disturbed. Later on in this book we will go into greater detail on how to locate the best times for associative remote viewing using solar weather, but first let's do a quick re-cap on solar weather conditions and ARV.

The best conditions for RetroPK and ARV sessions are as follows (from best to least) -

1 - Full Moon with no major solar flare activity occurring. Mild to low activity is preferred.

2 - Favourable solar Wind Speeds averaging 350.

3 - Quiet geomagnetic activity. Middle Latitude Fredericksburg K-indices below 11.

4 - The sun's F10.7cm radio flux. Rising or steady. Not necessary during full moons when above parameters exist.

When any of the above overlap with one another it creates a beneficial synergy.

ARV Cycle #1 - Solar Maximum / Increasing Solar Activity + First Quarter Moon = ARV sessions work best (RetroPK sessions may also show this effect, although further research is needed). Quiet to low geomagnetic activity. Solar wind speed around 350 and quiet to low geomagnetic activity are the key elements.

ARV Cycle #2 - Solar Minimum / Declining Solar Activity + Full Moon = ARV / RetroPK Sessions work best, unless solar flare activity is present. F10.7cm and Deviation from the Ecliptic are key elements. Above average geomagnetic activity is highly disadvantageous to ARV sessions during **ARV Cycle #2**.

As solar minimum approaches, ARV sessions should be conducted around full moons. As solar maximum approaches, ARV sessions should be conducted around the first quarter moon. The above variations are wholly dependent upon favourable solar wind speeds and quiet / low geomagnetic activity. Further research in the coming years should validate this hypothesis.

Forecasted   Sunspot   Cycles   until   the   Year   2140

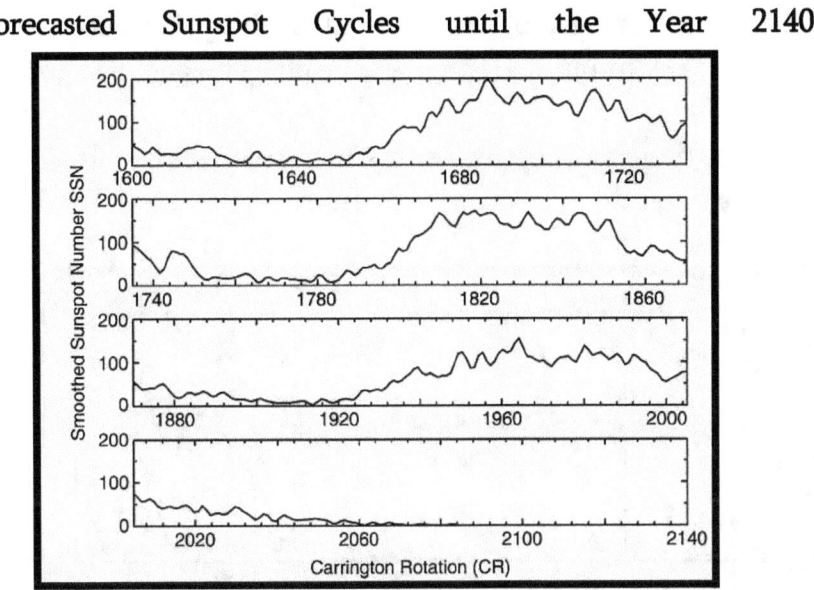

As can be seen in the above image, sunspots are forecast to decline for the next 50 years. This means we can continue to see major increases in cosmic rays over the next few decades.

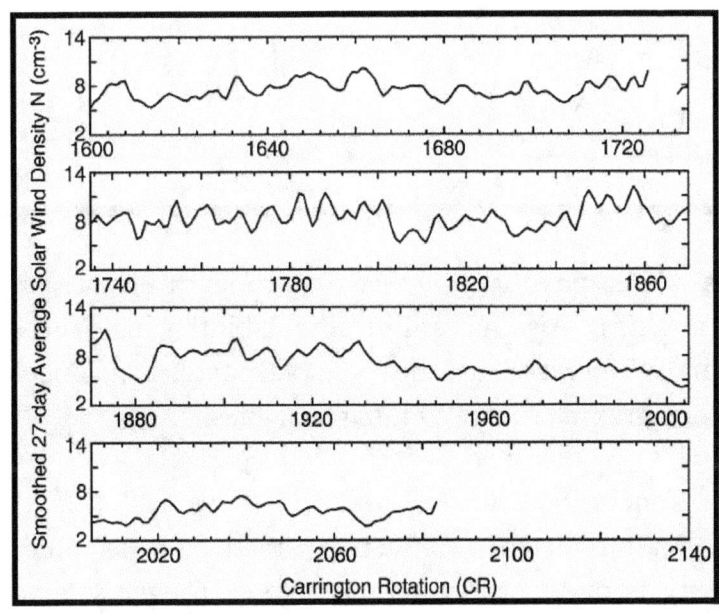

Above - Solar Wind Speed Forecast until the Year 2060

Images are courtesy of the paper titled: How Unprecedented A Solar Minimum. C. T. Russell et al. Feb 2010.

The last time sunspots were this low was between 1640 and 1700 and again from early 1800.

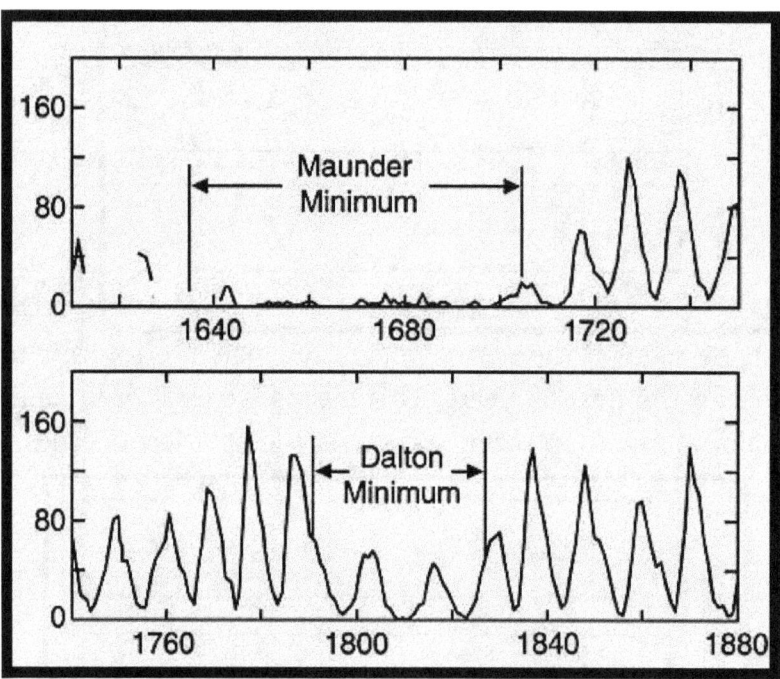

This would mean it is a great time to learn how to use one's PSI abilities. Also natural intuitive abilities may also start to naturally increase in the general population due to the aforementioned solar and lunar variables.

Final Conclusion

The reason RetroPK effects increase during quiet solar activity is due to more cosmic rays. Stronger solar activity causes less cosmic rays and if this occurs during full moons it can greatly hinder psychic ability. Now that we have covered much of the technical data of environmental factors that affect PSI functioning, let's next look at one of the prime keys to ARV success which is coherence. How does one use coherence to enhance PSI abilities? This was once a great mystery, but today we have the technology to uncover the answers.

References. Chapter 5.

1. Global solar wind variations over the last four centuries. M. J. Owens. Nov 2016.

2. The solar wind at solar maximum: comparisons of EISCAT IPS and in situ observations. A. R. Breen et al. July 2002.

3. Seasonal solar wind speeds for the last 100 years. Kalevi Mursula et al. Dec 2016.

4. Solar Periodic full moon effect in the fourmilab retropsychokensis project. Eckhard Etzold. pgs 249 to 251.

5. Chauvenet, William (1906). A Manual of Spherical and Practical Astronomy. I. J.B. Lippincott Co., Philadelphia Art. 365–367, p. 694–695, at Google books. 6. A Mixture of Ancient and Modern Understanding Concerning the Distance and Motion of the Moon. Kevin Krisciunas.

7. Mutagenesis and Background Neutron Radiation. Augusto Gonzalez. June 2014.

8. Long-Term Study of Heart Rate Variability Responses to Changes in the Solar and Geomagnetic Environment. Abdullah Alabdulgader et al. Feb 2018.

9. The Solar Cycle. David H. Hathaway. Sept 2015.

10. Effect of cosmic rays on computer memories. Ziegler JF and Lanford WA. Nov 1979.

11. Mutagenesis and Background Neutron Radiation. Augusto Gonzalez. June 2014.

12. Tom Simonite, Should every computer chip have a cosmic ray detector?, New Scientist, March 2008.

13. MIT Technology Review. Cosmic Rays, Neutrons And The Mutation Rate In Evolution. July 4th, 2014.

14. Soft error. Wikipedia.

15. Wilkinson, J.D. Bounds, C. Brown, T. Gerbi, B.J. Peltier,

J. (2005). "Cancer-radiotherapy equipment as a cause of soft errors in electronic equipment". IEEE Transactions on Device and Materials Reliability. 5 (3): 449–451. doi:10.1109/TDMR.2005.858342.

16. Franco, L., Gómez, F., Iglesias, A., Pardo, J., Pazos, A., Pena, J., Zapata, M., SEUs on commercial SRAM induced by low energy neutrons produced at a clinical linac facility, RADECS Proceedings, Sept. 2005.

17. Observation of seasonal variation of atmospheric multiple-muon events in the MINOS Near and Far Detectors. P. Adamson. et al. March 2015.

18. Annual Variations of the Galactic Cosmic Ray Intensity and Seasonal Distribution of the Cloudless Days and Cloudless Nights in Abastumani. M. V. Alania. et al. Sept 2015.

19. Seasonal Variation of the Underground Cosmic Muon Flux Observed at Daya Bay. The Daya Bay Collaboration. Jan 2018.

20. Day night variation of cosmic rays intensity at sea level under the influence of metrological fronts and troughs. HM Mok and KM Cheng.

21. Long-term occurrence probabilities of intense geomagnetic storm events. K. Tsubouchi and Y. Omura. Aug 2007.

22. Solar cycle distribution of major geomagnetic storms. Gui-Ming Le. June 2013.

23. Wang and Luhr. 2007. Tanskanen et al. 2011.

24. A 27 day persistence model of near-Earth solar wind conditions: A long lead-time forecast and a benchmark for dynamical models. M. J. Owens. et al March 2013.

25. Forecasting the peak of the present solar activity cycle. Author links open overlay panel. R.H. Hamid and B.A.

Marzouk. March 2018.

26. Near-earth solar wind flows and related geomagnetic activity during more than four solar cycles (1963–2011). Ian G. Richardson. April 2012.

27. New perspectives on contributing factors to the monthly behaviour of the aa geomagnetic index. Blanca Mendoza, et al. Dec 2016.

28. Influence of the Moon on the Earth's Magnetosphere at Various Phases of a Solar Activity Cycle. L. A. Akimov and N. P. Dyatel. Feb 2012.

29. Seasonal Variation of High-Latitude Geomagnetic Activity in Individual Years. E. I. Tanskanen. Sept 2017.

30. Seasonal Variation of High-Latitude Geomagnetic Activity in Individual Years. E. I. Tanskanen. Sept 2017.

31. Lethal manifestations of meteorological and cosmic factors. Guliaeva TL. Oct 1998.

Chapter 6. Electrical Activity of the Heart Surpasses that of the Brain

Your Heart Puts out More Energy than Your Brain

Because our thoughts pervade our waking consciousness, it can be easy for anyone to think that the brain emits the most energy out of all organs in the body. But did you know it really is your heart that is more powerful, electrical wise, than your brain? Your heart is the most powerful source of electromagnetic energy in your body. It produces the largest rhythmic electromagnetic field out of all the body's major organs. This

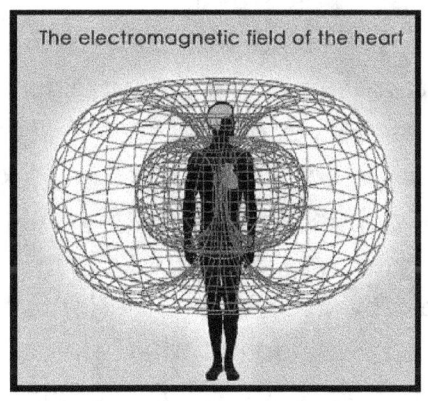

The electromagnetic field of the heart

electrical field is approximately 60 times stronger in its amplitude than the electrical activity generated by your brain.

Additionally the magnetic field that is emitted by your heart is 100 times greater in strength than the magnetic field generated by your brain. This magnetic field can be detected up to a distance of 3 feet away from your body using SQUID-based magnetometers. This is an interesting finding because it shows that emotions may be electromagnetic in nature. Hence the emotions we experience come from our heart, not our brain, which is why our emotions feel overwhelming and out of control at times due to this large electrical field emitted by our heart. However with self-regulation techniques that help one attain self-mastery, we no longer have to let this power get out of control and overwhelm us.

Early pioneers in psychophysiology were John and Beatrice Lacey. They specifically examined interactions occurring between the brain and heart. Throughout their 20 years of studies throughout the 1960s and 1970s, they found that our heart communicates with our brain and that this communication greatly affects our perception and how one reacts to the world.

Research by the Laceys confirmed a relationship between heart activity and cognitive performance. They investigated a person's performance on reaction time tasks that involved sensory intakes. Their studies found that **deceleration** in the participant's heart rate would occur during anticipatory periods **preceding** tasks and that it was associated with improved cognitive performance, or **faster reaction** times. Also they found that an acceleration in the heart rate contributed to reduced cognitive performance, or a slower reaction time (Ostir et al., 2001). Hence people with high blood pressure may have lower cognitive functioning (High blood pressure is linked to cognitive decline. June 2016.. /www.nia.nih.gov). (Ostir et al., 2001) also discovered that the larger a person's heart rate deceleration, the faster their reaction time (Lacey & Lacey, 1964, 1970, 1974). Other research has shown that changes in a person's cardiac field affects the growth rates of cells in culture (McCraty et al., 1998).

Additional research by the Laceys looking at activity occurring within single cardiac cycles, found that **cardiovascular activity influenced a person's perception** and their cognitive abilities. Inconsistencies did show up in their final results, however these inconsistencies were resolved by

German researchers Velden and Wölk who demonstrated cognitive performance had a rhythm consisting at approximately 10 hertz throughout a person's cardiac cycle.

Summary

The human heart is sending more information to the brain than the brain is sending to the heart. It may be that information is encoded and then communicated in the intervals between heartbeats. Let's next explore this amazing connection in greater detail in the next chapter.

Chapter 7. Coherence and the Heart

Why Feelings Generated by the Heart can be sensed by Others

Anyone knows the emotions they experience while watching a person cry or exude with joy. These emotions are due to the electromagnetic field generated by the heart. Research studies have found that when a person is in heart coherence, their heart is radiating a coherent electromagnetic signal into the nearby surrounding environment which is able to be felt by nearby animals, plants or the nervous systems of people nearby [1].

Definition of Coherence

It can be easy to think coherence only exists in a specific place at specific times, but in fact coherence can exist in synergy with multiple systems all at the same time. This is known as **cross-coherence**. As coherence increases in a system, which is coupled to nearby systems, it begins pulling the nearby systems into increased synchronization, resulting in increased functioning. Physiology defines cross-coherence when two or more of the human body's oscillatory systems, such as heart rhythms and respiration, become entrained with each other and begin operating at a single common frequency.

As an example, regular entrainment and frequency pulling exist between the body's respiratory, heart and blood pressure rhythms, very-low-frequency brain rhythms and electrical potentials as measured across the skin and craniosacral rhythms. When a person is in a coherent state of being, their heart pulls these biological oscillators into resonance with each other. This is due to the fact that coherence generates positive emotions which increases synchronization of the body's systems. This creates an increase of natural energy which enables one to function with greater effectiveness and efficiency.

Positive Emotions, Sine Waves and EEG Activity

A coherent heart rhythm is a rhythm that is a relatively harmonic, sine-wave like signal. This sine-wave like display has a very narrow, high-amplitude peak which exists in the low frequency (LF) region of the HRV power spectrum. It exhibits no major peaks in the high-frequency (HF) or very-low-frequency (VLF) regions. Sine waves look very similar to standing waves which was covered earlier on in this book. The image below are sine waves.

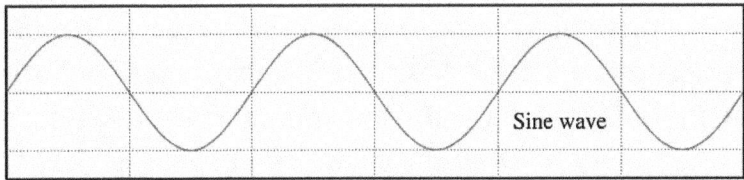
Sine wave

What is LF/HF?

The autonomic nervous system (ANS) consists of dual operating systems. These are the parasympathetic and sympathetic nervous systems which are reflected in the low- (LF) and high-frequency (HF) bands in heart rate variability.

When a person experiences positive emotions and their brainwave activity is wired to an electronic instrument to measure their brainwaves (EEG activity), their brainwaves display a sine-wavelike pattern on the computer screen. This sine-wave pattern comes from frequencies found in the heart's rhythm [2] [3]. As discussed earlier, standing waves are created in earth's magnetosphere during full moons. It may be that these coherent sine waves generated during HeartMath synergize with standing waves that have more intensity when the moon is full.

The dictionary definition of coherence is how stable or constant a logically integrated system is while in a coherent state [4]. A similar definition is the orderly, logical and aesthetically healthy relationship among parts [4]. Coherence implies consistency, correlations, connectedness and efficient energy utilization. Hence coherence means global order and wholeness where the whole is greater than the sum of its individual parts.

Geomagnetic Storms and EEG Alpha Brainwaves

What if our bodies were sensitive to changes in earth's magnetic field, wouldn't that be interesting because insects and other animals already use earth's magnetic field for navigation. Today we have instruments sensitive enough to answer this question. Doronin et al. examined a group of people and the effect of geomagnetic activity on their body. The study looked at the people's heart rate, blood pressure, EEG patterns and reaction times [5]. The study found that oscillations in earth's magnetic field showed identical periods taking place in the monitored EEG alpha rhythm. This study confirms geomagnetic activity affects the body's heart and brain which in turn exerts whole-body changes.

Further Reading

Coherence: A Novel Nonpharmacological Modality for Lowering Blood Pressure in Hypertensive Patients. Abdullah

Alabdulgader. May 2012.

Primary causes of enhanced geomagnetic activity

There are 2 primarily conditions contributing to a disturbed magnetic field upon our earth.

1 - **Coronal mass ejections**. These contribute to an increase in sunspots (Juan M. Borrero and Kiyoshi Ichimoto, 2011).

2 - **High speed solar winds**. These are caused by solar coronal holes (Feynman, 1982).

To put it simply, solar activity creates high speed solar winds which in turn affect earth's geomagnetic activity.

Coherence in Physics

This is a major and important part. As many of the devices, tools and parts used in the ARV amplification device exhibit strong coherence. For example the material Yttrium, used in our ARV device, is also used is lasers. Lasers are one of the strongest sources of coherent light.

Physics defines coherence as the degree and coupling of synchronization between independent oscillating systems. In specific instances when two or more oscillatory systems are operating at the same basic frequency, they become phase-or frequency-locked. An example of this is the oscillating photons in a laser [6]. This specific type of coherence is known as cross-coherence. It is this coherence that many scientists describe when they use the term 'coherence' in laser physics.

Coherence in Human Physiology

In this definition, cross-coherence takes place when two or more of a person's oscillatory systems, such as their heart rhythm and respiration, become entrained and begin operating at the same basic frequency. Coherence is highly beneficial to a person's parasympathetic nervous system and research is starting to show it may exhibit anti-aging and regeneration effects.

Heart Coherence

This type of coherence (also called cardiac coherence) is measured by HRV analysis where the pattern of a person's heart rhythm is ordered and sine wavelike. This takes place at the frequency of approximately 0.1 hertz (10 seconds). Another interesting fact is tensing one's large muscles in the legs (as occurs in sports such as soccer) in a rhythmical fashion at the rate of a 10-second rhythm induces coherent heart rhythms [7].

During our normal day to day lives our heart goes through various levels of coherence. If we were to look at heart coherence on a television screen using special equipment while performing an emotionally stressful task, it most likely would show incoherent waveforms.

Coherence Causes Increased Parasympathetic Nervous System Activity

It is the goal of many of us to have things fall into place with the least amount of effort. But life does not always work out this way. However, what if one could utilize the power of coherence to form coherent standing waves that form a ballet of frequencies which enhance one's intuitive abilities, thus avoiding future stress or even tragedy? The key to this is the **parasympathetic nervous system**.

A person who shifts into a coherent state causes their autonomic nervous system to move towards greater parasympathetic activity (**vagal tone**). This is the result of increased entrainment between diverse physiological systems, especially heart-brain synchronization. It is during this time that the body's main physiological systems are functioning with a high degree of harmony and efficiency and natural regenerative processes begin taking place.

Speaking from personal experience, I have found that after an especially hard workout, when I practice the HeartMath quick coherence exercise before going to bed, I feel much stronger and less sore the following morning, compared to if I hadn't practiced HeartMath.

The parasympathetic nervous system can also be stimulated by eating specific foods. For example a research study found that eating just a single dose of dark chocolate (10 grams) increased the parasympathetic nervous system. The study also found that significant increases were found in the person's heart rate variability [7b]. As we shall go into greater detail later on, increased heart rate variability (HRV) is good for remote viewing. This is an amazing study because these effects increase life expectancy because reductions in heart rate variability are associated with higher mortality including increased susceptibility to cardiovascular diseases (Reduced heart rate variability and mortality risk in an elderly cohort. The Framingham Heart Study. Tsuji H et al. Aug 1994).

Recent research studies by the HeartMath Institute show that the electromagnetic field of a person's heart can be detected by the nervous systems of other people or nearby animals without them being in direct physical contact with one another [8]. Perhaps the first person to open a cacao ice cream shop might find it highly successful as the effects of increased HRV as people eat cacao would spread to nearby people drawing them in like bees to honey.!

Essential Oils that Stimulate the Parasympathetic Nervous System

As explained at the beginning of this book, essential oils were one of the main reasons enhanced success of remote viewing was made possible. Peer review published journals don't hold all the answers. There are some good independent researchers doing some very good work. While research studies are ongoing studying the role the parasympathetic nervous system plays in the body, one independent researcher found that a mixture of clove and lime essential oils worked extremely well at creating a stronger / slightly elevated parasympathetic nervous system. The researcher also found that the combination greatly reduced his headaches [9].

A recently published study found that stimulating the vagal pathways (located just below the ear lobes) caused significant reductions in migraine and cluster headaches (Mauskop, 2005). The best ratio is 3 drops each of lime and clove essential oil and one drop of copaiba essential oil to amplify the therapeutic benefits [10]. Next apply just a tiny drop just below each ear lobe.

Further Reading

Copaiba Oil-Resin Treatment Is Neuroprotective and Reduces Neutrophil Recruitment and Microglia Activation after Motor

Cortex Excitotoxic Injury. Adriano Guimarães-Santos. et al.
Feb 2012.

It is my theory that the autonomic nervous system plays a
major role in aging and substances that affect the autonomic
nervous system may have profound effects on slowing down
or even reversing aging. For example a study titled:
Autonomic Nervous System Network And Liver
Regeneration, that was published by Kenya Kamimura and
colleagues in April 2018, states that a relationship exists
between the body's autonomic neural network and the ability
of the liver to regenerate via inter-organ communication.
This sounds a lot like cross-coherence It is also interesting to
note that clove has been shown to be of great benefit to
people suffering from liver cirrhosis [10b].

Tibetan monks regard the spice Clove as a very, very
special spice. Clove has the highest polyphenol content out of
any known food and as we discussed earlier, foods with a
high polyphenol content protect against radiation. It is also
interesting to note that a 2015 study showed that eugenol,
which is the primary ingredient of Clove, allows the body to
maintain normal gastrointestinal motility even during stress.
The study theorizes that eugenol activates stress-responsive
regions of the brain, causing them to produce stress-response
hormones that are released throughout the body, thus
enhancing one's ability to rapidly adapt to stress [11].

One interesting point I would like to point out here is that
when solar weather conditions are extremely favourable,
which the Solar Institute has termed a "**condition green**" [12]
that it is one of the very best times to get a massage or to do
exercises that relieve stress and tension. It may be that

during these condition green periods that the body is 're-set' into health, vitality and well being and that certain exercises as well as essential oils combined with one's intent help "push through" into a new cycle of well being and health. This needs further research to verify, as this observation is only based on years of observational research.

Another very interesting discovery I had noted is that the human parasympathetic nervous system regulates the body's bladder and colon [13] [14]. Our ARV sessions are always conducted around midnight, as that has been the time we have always achieved the very best results. In Traditional Chinese Medicine, the Chinese Organ Body Clock shows that **between 11 p.m. to 1 a.m.** the energy in the body's **Gall Bladder Peaks**. It is when the energy of yang begins growing.

Hourly QI Flow Through Organs		
Time Period	Organ	Month
11 a.m. to 1 p.m.	Heart	June
1 p.m. to 3 p.m.	Intestines	July
3 p.m. to 5 p.m.	Bladder	August
5 p.m. to 7 p.m.	Kidneys	September
7 p.m. to 9 p.m.	Pericardi-um	October
9 p.m. to 11 p.m.	Triple Burner	November
11 p.m. to 1 a.m.	**Gall Bladder**	December
1 a.m. to 3 a.m.	Liver	January

3 a.m. to 5 a.m.	Lungs	February
5 a.m. to 7 a.m.	Colon	March
7 a.m. to 9 a.m.	Stomach	April
9 a.m. to 11 a.m.	Spleen	May

Parasympathomimetic Substances

Parasympathomimetic substances, are also known as cholinomimetics. These stimulate the parasympathetic nervous system and are usually only available by prescription [15]. They are also sometimes called cholinergics due to the fact that acetylcholine is the neurotransmitter used by the parasympathetic nervous system. These types of substances act by stimulating the nicotinic or muscarinic receptors (thus mimicking the effects of acetylcholine), or they can have an effect indirectly by inhibiting cholinesterase, which also enhances the release of acetylcholine release, or via other methods [16]. Bethanechol for example stimulates nicotinic or muscarinic receptors and is used to treat urinary retention resulting from diabetic neuropathy of the bladder or general anaesthetic [17].

Pilocarpine is another parasympathetic nervous system stimulant which is used to treat increased pressure inside the eye and dry mouth [18]. It is also used for reducing the glare at night from lights [19]. Pilocarpine can be extracted from the plant jaborandi (pilocarpus microphyllus stapf) [20]. Also as a side note, many plants that treat diabetes also have potent anti-aging properties. This could be why jaborandi is used to treat diabetes [21]. The book: A Treatise on Bright's Disease and Diabetes: With Especial Reference by James Tyson, George Edmund De Schweinitz, states to pour 4 ounce of hot water over Jaborandi leaves for excellent results in treating diabetics [22] [23].

Oxytocin. The Bonding Molecule

Further research on the heart has found that a person's heart has cells which synthesize and release the neurotransmitters catecholamines (norepinephrine, epinephrine and dopamine). These neuro-transmitters were originally thought to be produced by neurons in the human brain and ganglia [24]. Further research found that the heart manufactures and secretes the substance oxytocin. Oxytocin acts as a neurotransmitter and is referred to as a love or social\bonding hormone. This is a rather interesting finding because love can be partly defined as a coherent form of emotion. Oxytocin has a strong presence in lactation and childbirth. It also plays a role in tolerance, cognition, trust, friendship and the establishment of enduring pair-bonds. What is most interesting is the amount of oxytocin produced in the human heart is also about the same amount as produced in the human brain [25].

How to Increase Oxytocin levels

Oxytocin can be released by various types of non-noxious sensory stimulation, for example by touch and warmth, i.e., massage. Ingestion of specific foods trigger a release of oxytocin by activation of vagal afferents. Most likely, oxytocin can also be released by stimulation of other senses such as from specific odours and fragrances (olfaction) as well as by sound and light (external stimuli). Also psychological mechanisms trigger oxytocin (positive interactions such as touch and psychological support). This could be where the healing effects of touch come from.

Singing Increases the Social Bonding Hormone Oxytocin

A research study found that a person's oxytocin levels increased in response to improvised singing (also called scat singing). The study theorized that the higher amounts of oxytocin may have been due to the social effects of improvising musically with others [26]. Also people who were about to undergo surgery that listened to relaxing music showed an increase in their oxytocin levels and reported less anxiety compared to the control group that listened to no music (Nilsson, 2009). Also singing in a choir has been shown to increase salivary oxytocin as well as generate positive emotional states (Kreutz et al., 2004 Kreutz, 2014). In a study involving amateur and professional singers, the study found that the singer's peripheral oxytocin increased after singing for 45 minutes. (Grape et al., 2002). The same study also found that the amateur singers showed a decrease in post-singing levels of cortisol, whereas the professional singers exhibited the opposite. Hence showing higher levels of perceived stress and arousal in the singers who were professionals. This same trend is consistent with Fancourt et al. (2015), who found that low-stress singing with nobody listening reduced levels of salivary cortisol and cortisone, whereas high-stress singing in front of large audiences increased a person's level of glucocorticoids. Hence, depending on a person's singing experience, it is possible that singing can be stressful for some. The study concluded that oxytocin plays a major role in the health and social benefits of music [27]. This is a rather interesting study because the same effects may also occur for people who do public speaking. There are some people who enjoy public speaking and others who are terrified. Hence future studies may find that professional public speakers exhibit higher oxytocin levels and people who have never spoken before a live audience would exhibit lower oxytocin levels.

Oxytocin as a Natural Fear Repellent

The purpose of the brain's amygdala is to process fear and anxiety. It detects threats and links the threats to various defensive behaviours. This is accomplished by connections within the central nucleus of the amygdala to the stem region of the brain and also to hypothalamic regions which are responsible for organizing fear responses. The role oxytocin plays is it tempers the electrical activity that flows into the brain's amygdala. Hence oxytocin has been shown to suppresses activity in the brain's amygdala [28]. In summary, the amygdala processes fear and communicates it to the rest of the brain [29]. A research study involving 18 volunteers inhaled either oxytocin or a placebo before performing a task in which they sorted pictures of angry or fearful faces and threatening-life scenes. During this test researchers monitored the subjects' brain wave activity using magnetic resonance imaging (MRI) and discovered that the group that had inhaled oxytocin had reduced activity in the amygdala part of their brain. The study concluded that inhalation of oxytocin can be of benefit to those who suffer from anxiety related disorders (Oxytocin Attenuates Amygdala Reactivity to Fear in Generalized Social Anxiety Disorder. Izelle Labuschagne.et al. Aug 2010).

Another study found that newborn macaques that inhaled oxytocin showed increased positive social behaviour [29].

Herbs with oxytocic properties - Bidens Pilosa L. (Asteraceae) Leaf. **Reference**: In Vivo and In Vitro Effects of Bidens Pilosa L. (Asteraceae) Leaf Aqueous and Ethanol Extracts on Primed-Oestrogenized Rat Uterine MuscleLongo Frida et al. Oct 2007.

Fruit of Hunteria umbellate. **Reference**: Phytochemical components of Hunteria umbellata (K. Schum) and its effect on isolated non-pregnant rat uterus in oestrus. Falodun A et al. July 2006.

Essential Oils and their Effects Upon the Heart
It is our hypothesis at the Solar Institute that specific substances exist that can increase the amount of energy the heart can hold. In other words these substances allow the energy flow that accompanies the energy of the heart to flow smoother resulting in an increase. This is much like a capacitor that has stacks of separate layers, allowing it to hold large amounts of energy (enhancement of an energetic amplitude). This dramatically enhances the sensitivity of information received. Let's first take a look the substance limonene which we briefly covered earlier.

The Power of Limonene
Limonene is a hydrophobic (repels moisture) terpene and the word originates from lemon peels, due to lemon peels containing vast amounts of Limonene. A research study

conducted in June of 2015 looked at what main compounds were present in the substances lemon, tangerine and sweet orange. The study found that the main compound in each of the oils was the terpene limonene [30]. Lemon had the largest amounts followed by tangerine and sweet orange. During our ARV sessions we slice open a lemon and put it on the table and add a few drops of Jasmine Essential Oil. This seems to greatly enhance the success of the ARV session. Tangerine also contains an abundance of the terpenoids a-pinene and ß-pinene.

Because a lot of stress can be put on the heart during an ARV session due to the intensified mental concentration and effort, and excess stress can cause illness, the limonene may also be strengthening the heart as well as warding off bad bacteria as limonene is a powerful germ fighter. A research study found that limonene, especially when heated, was one of the most effective treatments for killing E. Coli when limonene was used for food preservation [31].

Yuzu essential oil and Limonene

A research study found that yuzu (Citrus junos Sieb. ex Tanaka) contains high amounts of limonene. The major components were as follows: limonene 78.02 %, terpinene 9.32 %, myrcene 1.77 %, a-pinene 1.34 %, and linalool 0.03 %.

Reference

Aromatic effects of a Japanese citrus fruit—yuzu (Citrus junos Sieb. ex Tanaka)—on psychoemotional states and autonomic nervous system activity during the menstrual cycle: a single-blind randomized controlled crossover study Tamaki

Matsumoto. April 2016.

Today food grade limonene can be bought online and is commonly used as a preservative and sweetener when a citrus taste is desired.

Limonene and White Blood Cells

A study looking at Balb/c mice found that when they were given carvone (a substance found in essential oils) that their white blood cells became stronger on the 12th day. However when they were given limonene it only took 9 days for their immune systems to reach maximum strength. The study also found that the terpenoids (also found in essential oils) increased their antibodies, enhancing the strength of their spleen and bone marrow [32] [33].

Summary

Limonene may be enhancing the strength of the body's physiological systems by relaxing the nervous system, which in turn strengthens one's cognitive abilities.

Limonene in Nature

A study found that limonene in Citrus reached its maximum concentration during the mature stage with the highest amounts in the peels [34].

Essential Oils containing Limonene [35] [36] [37]

Orange
Grapefruit
Mandarin

Japanese Catnip
Spearmint
Eucalyptus
Lemon
Tanagerine
Bergamot (also contains linalool and linalyl acetate)
Caraway (also contains carvone)
Celery Seed (74.6%)

A research study using rats found that after breathing in grapefruit oil (which contains an abundance of limonene) that it stimulated their sympathetic nerves due to its ability to activate histamine H1 receptors. The same study found limonene alone caused a similar response (Olfactory stimulation with scent of grapefruit oil affects autonomic nerves, lipolysis and appetite in rats. Shen J et al. June 2005).

What is a H1 Receptor?
A H1 receptor is a type of biological mechanism that helps the body fight allergic reactions.

Also bergamot essential oil enriched with Limonene has been shown to exhibit vaso-relaxant effects [38]. Vasor-elaxant effects are a beneficial reaction for the heart. From personal experience, inhaling bergamot essential oil a few hours before an ARV session helps the body relax. It is also interesting to note that limonene will dissolve fish odour [39].

Spices containing an abundance of Limonene [40]

Dill
Parsley
Coriander
Mint
Peppermint

Essential Oils with an abundance of Limonene [41]

Peppermint - D-carvone (58.79 %) limonene (28.29 %)
mint essential oil. The highest amounts were found when the
peppermint was harvested in December.

Foods high in Limonene

C.latifolia Tanaka (Persian Limes) [41]

Limonene is a powerful natural Antimicrobial

A research study found that essential oils of C. maxima and C.
sinensis in a ratio of 1:1 showed powerful fungitoxic
properties allowing it to fight mould. The combination also
inhibited aflatoxin and the limonene in the essential oils
showed powerful antiafla-toxigenic activity. Both of the
essential oils also showed antioxidant activity in a dose
dependent manner. The study concluded that these essential
oils may be a good for extending the shelf life of food by
reducing aflatoxin production, fungal infestation and lipid
peroxidation [42].

The spice Fennel also contains limonene and a-pinene [43]
and is also used in the longevity formula known as St.
Germain. This is a rather interesting finding because the St.
Germain Formula, invented by Alchemist St. Germain,

contains Fennel Seed and the formula was supposedly used by him who was reputed to have lived to well over 100 years.

Limonene also has extremely powerful anti-cancer properties and up to 21% Limonene is found in the essential oil of Pinus Koraiensis, which is used as a folk remedy for heart disease [44]. Limonene also reduces bio-film formation of Streptococcus pyogenes between 75% and 95% against specific pathogens. This shows that limonene reduces the ability of bad bacteria to adhere to surfaces, making it a powerful germ killer [45]. A research study found that limonene, when given in low amounts to fruit flies, extended their lifespan and protected their DNA against free radicals [46].

Limonene and Stress

A research study found that Limonene has been shown to protect against the systolic increase in blood pressure and attenuate hypertension in animal studies [47]. Also limonene has been shown to greatly enhance the ability of rats to distinguish the speed at which they recognize familiar objects during memory tests. Also limonene reduced the time spent by animals in a maze. The study also found that limonene significantly enhanced the grip-strength in rats that had just had a stroke. This is a major finding because grip strength is also an indicator of aging. The stronger the grip-strength the healthy the older person. The study concluded that limonene protected against ischemia-induced memory and the impairment of cognitive abilities while they experienced hypertension and these effects were attributed to Limonene's ability to causes changes in dopamine levels [47].

Summary

Limonene's ability to extend lifespan and strengthen cognitive abilities may be due to its anti-fungal / antibacterial properties. Hence could the main reason neurocognitive ability declines as one ages be due to micro-bacteria and that this micro-bacteria can be reduced by the substance Limonene?

Another study looked at the effects of inhaling jasmine tea on autonomic nerve activity and moods in healthy volunteers. The study found that jasmine tea and lavender odors caused significant decreases in the volunteers' heart rate, producing calm and vigorous mood states. The study concluded that jasmine tea, when inhaled at low levels, exhibited sedative effects on the body's autonomic nervous system and moods and that linalool mimics these effects. It is important to note here that the study stated that just the right amount was used, as overuse can cause the beneficial effects to be lessened [48]. This could also be why a research study found that lavender, which contains 40% linalool, showed that mice who received lavender showed significantly improved cognitive abilities. The study concluded that the reason for the strengthened cognition was due to the lavender alleviating oxidative stress and by enhancing weakened synaptic plasticity. The study concluded that lavender essential oil and its main component linalool, has the potential to prevent or improve cognitive deficits in mice that have Alzheimer's [49].

What is also interesting is lavender has been shown to excite the parasympathetic gastric nerve in the body's nervous system [50] and that activity of the parasympathetic nervous system becomes reduced as one ages [51]. A research study found that inhaling lavender had a short-term effect on HRV and also increased parasympathetic modulation in the nervous system [52].

The substance linalool and eugenol exhibit synergy with one another. Hence this combination has been shown in research studies to kill Leishmania (L.) amazonensis and Trypanosoma cruzi at low doses. This was shown in a research study that found that when linalool was mixed with eugenol that it significantly decreased the number of parasites in peritoneal mouse macrophages [53].

Linalool and the Parasympathetic Nervous System

Research studies have found that inhaling lavender has persistent short-term effects on HRV and that it also leads to an increase in the parasympathetic nervous system [54]. This effect is most likely due to the substance linalool, which is also found in the Jasmine Flower [55]. Studies have found that linalool produces vigorous and calm moods as well as elicits an increase in HF (high-frequency) and a significant decrease in heart rate [56]. Linalool is quickly absorbed into the blood stream. Bradley et al. found after oral administration of lavender essential oil, that the linalool was detectable in the blood stream within 10 min. This than peaked after 30 min whereupon it was no longer detectable 45 minutes later [57].

Other independent studies have found that linalool produces vigorous and calm moods as well as elicits an increase in HF (high-frequency) and a significant decrease in heart rate [57].

Further Reading

Sedative effects of the jasmine tea odour and linalool, one of its major odour components, on autonomic nerve activity and mood states.   Kuroda K, Inoue N, Ito Y, et al.   European Journal of Applied Physiology. 200595(2-3):107–114.

Linalool Enhances the Ability to Solve Math Problems

A research study involved human subjects evaluated in three different conditions (linalool aroma, no aroma and peppermint aroma) and used the PEBL (Psychology Experiment Building Language) Math Processing Task. This is a short test that involves answering math problems as accurately and as quickly as possible. The study found that the participant's exposed to the aroma of linalool exhibited increased accuracy compared to their peers who received no linalool aroma. The Peppermint group who inhaled the peppermint aroma showed no effect on math accuracy or response time compared to the control group that received no aroma. The study concluded that inhaling linalool can greatly increase mathematical accuracy and / or general task performance [58].

Another research study found linalool (100 mg/kg, i.p.) significantly improved the cognitive performance model in mice who underwent the Morris Water Maze Test and Step-Through Test. Linalool also effectively reversed Aß1–40 induced hippocampal cell injury. The study concluded that linalool attenuates cognitive deficits and that it has neurocognitive protection with the potential to be a neuroprotective substance for Alzheimer's therapy [59].

Solar Eclipses and Bacteria

From our many years of conducting ARV sessions, we have found an increase in accuracy during solar eclipses.   Hence the reason for this accuracy could be due to solar eclipses affecting bacteria. A study titled: Effect of solar eclipse on microbes conducted by Amrita Shriyan, Angri M. Bhat, and Narendra Nayak that was published in March 2011 found that a solar eclipse exhibited differences in the antibiotic sensitivity of microorganisms [60] and in a separate study found that the killing of bacteria was enhanced during a solar eclipse [61]. It would be interesting to do a study looking at the effects linalool has on microorganisms during a solar eclipse.

Alpha Brainwaves and Limonene

I mention in great detail about how alpha waves play a major role in remote viewing sessions. Hence an increase in alpha waves appears to increase the accuracy of ARV sessions. What effect does Limonene have on alpha brainwaves? Let's take a look at the data.

A research study conducted by Sowndhararajan et al. examined the effects of inhaling terpinolene and limonene and its effects on EEG brainwave activity. The study found that women responded well to both substances which caused a SIGNIFICANT increase in their fast (absolute) alpha brainwave activity. The study also found that inhaling terpinolene and limonene caused an increase in beta brainwaves in men, and that the beta brainwave activity was not as strong in women [62]. Let's now return to the antibacterial functions of terpenes. Maybe Pinene also has similar anti-bacterial effects. Let's take a look at the data.

Pinene and Cymene

The essential oil of Nigella sativa (Black Cumin) has been

found to contain an abundance of a-Pinene and p-cymene (monoterpenes). When rats and guinea-pigs were given Nigella sativa (4-32 µL/kg), researchers discovered that it decreased their arterial blood pressure (El Tahir et al., 1993 El Tahir & Ageel, 1994). In another study conducted by El Tahir et al. (2003) they found that an intravenous administration of a-pinene and p-cymene caused a decrease in the rats' arterial blood pressure and that it decreased their heart rate (El Tahir et al., 1993), (Menezes et al. (2010).

Another study found that ß- pinene induced endothelium- independent vaso-relaxation, which is another term for a relaxed and calm heart [63] .

Eugenol
Eugenol has been commonly used as a folk remedy for hypertension. Research studies have found that Eugenol reduces systemic blood pressure and relaxes conduit and ear arteries [64]. This is an interesting finding because the inner ear is sensitive to changes in pressure. And because our ARV sessions always do well when the barometric air pressure has peaked and is starting to fall, the inner ear may be acting as a type of tuning / antenna mechanism that helps in the overall ARV process.

What is Eugenol?
Eugenol is a phenylpropanoid which has a pale yellow color (although it can be colourless) with an aromatic scent. It has an oily feel and is extracted from the essential oils of basil, nutmeg, clove, cinnamon and bay leaf.

Eugenol Synergy

Eugenol synergizes with citral and thymol. These combinations are usually used to reduce on the growth of Crithidia fasciculata and Trypanosoma cruzi bacteria [65].

References. Chapter 7

1. McCraty R. The energetic heart: bio electromagnetic interactions within and between people. Boulder Creek (CA): HeartMath Research Center, Institute of HeartMath 2002

2. McCraty, R., et al., The effects of emotions on short-term power spectrum analysis of heart rate variability. Am J Cardiol, 1995. 76(14): p. 1089-93.

3. Tiller, W.A., R. McCraty, and M. Atkinson, Cardiac coherence: a new, non-invasive measure of autonomic nervous system order. Altern Ther Health Med, 1996. 2(1): p. 52-65

4. Kleiger, R.E., et al., Decreased heart rate variability and its association with increased mortality after acute myocardial infarction. American Journal of Cardiology, 1987. 59(4): p. 256-262

5. Doronin, V.N., Parfentev, V.A., Tleulin, S.Zh, .Namvar, R.A., Somsikov, V.M., Drobzhev, V.I. and Chemeris, A.V., Effect of variations of the geomagnetic field and solar activity on human physiological indicators. Biofizika, 1998. 43(4): p. 647-653.

6. Damasio, A., Looking for Spinoza: Joy, Sorrow, and the Feeling Brain. 2003, Orlando: Harcourt.

7. Lehrer, P., et al., Effects of rhythmical muscle tension at 0.1Hz on cardiovascular resonance and the baroreflex. Biol Psychol, 2009. 81(1): p. 24-30

7b. A single dose of dark chocolate increases parasympathetic modulation and heart rate variability in healthy subjects. Ana Amélia Machado DUARTE1 et al. Dec 2016.

8. McCraty, R., The energetic heart: Bioelectromagnetic communication within and between people, in Bioelectromagnetic Medicine, P.J. Rosch and M.S. Markov, Editors. 2004, Marcel Dekker: New York. p. 541-562.

9. The Creation of Parasympathetic Blendby Jodi Cohen | Apr 25, 2016

10. Copaiba. www.youngliving.com.

10b. Eugenol-rich Fraction of Syzygium aromaticum (Clove) Reverses Biochemical and Histopathological Changes in Liver Cirrhosis and Inhibits Hepatic Cell Proliferation. Shakir Ali et al. Dec 2014).

11. Protective effect of eugenol against restraint stress-induced gastrointestinal dysfunction: Potential use in irritable bowel syndrome. Garabadu D et al. July 2015

12. www.ez3dbiz.com/in_depth.html

13. Participation of the parasympathetic and sympathetic nerves in regulation of gallbladder motility in the dog. Yamasato T and Nakayama S. April 1990

14. The autonomic innervation of the liver and gallbladder of Rana ridibunda. Azanza MJ.

15. Dorlands Medical Dictionary:parasympathomimetic.

16. Brenner, G. M. (2000). Pharmacology. Philadelphia, PA: W.B. Saunders Company. ISBN 0-7216-7757-6.

17. Gorsky, Meir Epstein, Joel B. Parry, Jamie Epstein, Matthew S. Le, Nhu D. Silverman, Sol (2004-02-01). "The efficacy of pilocarpine and bethanechol upon saliva production in cancer patients with hyposalivation following radiation therapy". Oral Surgery, Oral Medicine, Oral Pathology, Oral Radiology, and Endodontics. 97 (2): 190.

18. Hamilton, Richart (2015). Tarascon Pocket Pharmacopoeia 2015 Deluxe Lab-Coat Edition. Jones & Bartlett Learning. pp. 254, 412. ISBN 9781284057560.

19. Zhang L, Weizer JS, Musch DC (2017). "Perioperative medications for preventing temporarily increased intraocular pressure after laser trabeculoplasty". Cochrane Database

20. Production of pilocarpine in callus of jaborandi (pilocarpus microphyllus stapf). Ilka Nacif De Abreu et al. Aug 2005.

21. Beneficial Use Of Jaborandi In Cases Of Diabetes Insipidus Or Polydipsia. J. M. Brown, M.B., And F. Alsop, M.B.). August 1875.

22. Beneficial Use Of Jaborandi In Cases Of Diabetes Insipidus Or Polydipsia. J. M. Brown, M.B., And F. Alsop, M.B.). August 1875.

23. A Treatise on Bright's Disease and Diabetes: With Especial Reference. By James Tyson, George Edmund De Schweinitz.

24. Huang, M., et al., Identification of novel catecholamine containing cells not associated with sympathetic neurons in cardiac muscle. Circulation, 1995. 92(8(Suppl)): p. I-59.

25. Gutkowska, J., et al., Oxytocin is a cardiovascular hormone. Brazilian Journal of Medical and Biological Research, 2000. 33: p. 625-633.

26. The neurochemistry and social flow of singing: bonding and oxytocinJason R. Keeler et al. Sept 2015.

27. The neurochemistry and social flow of singing: bonding and oxytocin. Jason R. Keeler. Sept 2015

28. Oxytocin reduces amygdala activity, increases social interactions and reduces anxiety-like behaviour irrespective of NMDAR antagonism. Rosanna Sobota et al. Aug 2015.

29. Inhaled oxytocin increases positive social behaviours in newborn macaques. Elizabeth A. Simpson. et al. April 2014.

30. Changes in the Composition of Aromatherapeutic Citrus Oils during Evaporation. George W. Francis and Yen Thuy Hoang Bui. June 2015.

31. Mechanism of Bacterial Inactivation by (+)-Limonene and Its Potential Use in Food Preservation Combined Processes. Laura Espina et al. Feb 2013.

32. Immunomodulatory activity of naturally occurring monoterpenes carvone, limonene, and perillic acid. Raphael TJ et al. May 2003

33. Citrus peel use is associated with reduced risk of squamous cell carcinoma of the skin. Hakim IA et al. 2000.

34. Changes of peel essential oil composition of four Tunisian citrus during fruit maturation. Bourgou S et al. May 2012.

35. Antioxidant Activities and Volatile Constituents of Various Essential Oils. Alfreda Wei, and Takayuki Shibamoto. Feb 2007.

36. Influence of Fragrances on Human Psychophysiological Activity: With Special Reference to Human Electroencephalographic Response. Sowndhararajan K and Kim S. Nov 2016.

37. Olfactory stimulation with scent of grapefruit oil affects autonomic nerves, lipolysis and appetite in rats. Shen J et al. June 2005.

38. The essential oil of Citrus bergamia Risso induces vasorelaxation of the mouse aorta by activating K(+) channels and inhibiting Ca(2+) influx. Kang P et al. May 013.

39. Co-Encapsulation of Fish Oil With Phytosterol Esters And Limonene. Chen, Qiong. The University of Auckland. 2012.

40. Volatile Composition of Essential Oils from Different Aromatic Herbs Grown in Mediterranean Regions of Spain. Hussein El-Zaeddi et al. May 2016.

41. Chemical profile, antifungal, antiaflatoxigenic and antioxidant activity of Citrus maxima Burm. and Citrus sinensis (L.) Osbeck essential oils and their cyclic monoterpene, DL-limonene. Singh P et al. June 2010.

42. Antioxidant and anticarcinogenic effects of methanolic extract and volatile oil of fennel seeds (Foeniculum vulgare). Mohamad RH et al. Sept 2011

43. Can Estragole in Fennel Seed Decoctions Really Be Considered a Danger for Human Health? A Fennel Safety UpdateL. Gori et al. July 2012.

44. Antioxidant and anticarcinogenic effects of methanolic extract and volatile oil of fennel seeds (Foeniculum vulgare).Mohamad RH et al. Sept 2011.

45. Limonene inhibits streptococcal biofilm formation by targeting surface-associated virulence factors. Subramenium GA et al. Aug 2015.

46. Role of citrus juices and distinctive components in the modulation of degenerative processes: genotoxicity, antigenotoxicity, cytotoxicity, and longevity in Drosophila.Fernández-Bedmar Z et al. 2011.

47. Protective effects of D-Limonene against transient cerebral ischemia in stroke-prone spontaneously hypertensive rats. Xifeng Wang et al. Nov 2017.

48. Stimulating effect of aromatherapy massage with jasmine oil. Hongratanaworakit T. Jan 2010.

49. The Protective Effect of Lavender Essential Oil and Its Main Component Linalool against the Cognitive Deficits Induced by D-Galactose and Aluminum Trichloride in MicePan Xu, et al. April 2017.

50. Lavender and the Nervous System. Peir Hossein Koulivand et al. March 2013.

51. Aging of the autonomic nervous system Shimazu T et al. June 2005.

52. The effect of lavender aromatherapy on autonomic nervous system in midlife women with insomnia. Chien LW et al. Au 2012.

53. Effects of linalool and eugenol on the survival of Leishmania (L.) infantum chagasi within macrophages. Dutra FL, et al. Acta Trop. 2016.

54. The Effect of Lavender Aromatherapy on Autonomic Nervous System in Midlife Women with Insomnia. Li-Wei Chien et al. Aug 2011.

55. Sedative effects of the jasmine tea odour and (R)-(-)-linalool, one of its major odour components, on autonomic nerve activity and mood states. Kuroda K et al. Oct 2005. peppermint and linalool (lavender).

56. Effects of orally administered lavender essential oil on responses to anxiety-provoking film clips. Bradley BF et al. June 2009.

57. Sedative effects of the jasmine tea odour and (R)-(-)-linalool, one of its major odour components, on autonomic

nerve activity and mood states. Kuroda K et al. Oct 2005.
peppermint and linalool (lavender)

58. The Effects of Linalool and Peppermint Aroma on
Cognitive Performance. Kaufman, Robert et al. May 2017.
The Ohio State University. Department of Psychology
Undergraduate Research Theses 2017.

59. Protective effects of linalool against amyloid beta-
induced cognitive deficits and damages in mice. Pan Xua et al.
Oct 2017.

60. Effect of solar eclipse on microbes. Amrita Shriyan et
al. Mar 2011.

61. Killing of bacteria during solar eclipse and its
biological implications. Banerjee SK ad Chatterjee SN. 1983.

62. Influence of Fragrances on Human
Psychophysiological Activity: With Special Reference to
Human Electroencephalographic Response. Sowndhararajan
K and Kim S. Nov 2016.

63. Characterization and Antihypertensive Effect of the
Complex of (-)-ß- pinene in ß-cyclodextrin. Moreira IJ,. 2016

64. Eugenol dilates rat cerebral arteries by inhibiting
smooth muscle cell voltage-dependent calcium channels.
Dieniffer Peixoto-Neves. Nov 2015.

65. Combination of the essential oil constituents citral,
eugenol and thymol enhance their inhibitory effect on
Crithidia fasciculata and Trypanosoma cruzi growth. Camila
M.O. et al. Oct 2015.

Chapter 8. Attaining Self-Mastery through Coherence. Secrets of Self Regulation.

Emotions can easily work two ways, they can bring us joy or pain. We go through life to try and avoid pain and experience its pleasures. Where and how a person focuses their attention has powerful effects on the information that gets processed at higher levels. For example, if a person is in a noisy room filled with numerous conversations they have an ability to tune out the unwanted noise and focus on a single conversation that is of interest to them. Hence, we can achieve self-mastery via self-regulation techniques to create feelings of gratification and satisfaction even in the midst of chaos. Hence failure to effectively achieve self-mastery and maintain control, results in feelings of impatience, anxiety, frustration, depression or hopelessness.

Self-Regulation Techniques

Used alone or in combination with HRV coherence biofeedback training, self-regulation techniques have been shown to accelerate recovery from stressors and trauma and greatly increase one's resilience [1] [2] [3] [4]. This is a major finding because it scientifically shows that self-mastery techniques can be used to enhance one's health, resilience and recover from trauma.

In order for one to achieve even a basic level of self-mastery, techniques that create brain / heart coherence are necessary. This is because as we covered earlier, the electrical activity of the heart amplifies emotions to such an extent that they can overpower the thinking part of the brain. These tools we are about to cover are the best tools because they have undergone rigorous scientific study and are proven effective. To put it simply, using a coherence tool gives one the keys to self-regulation, which in turn allows one to achieve   self-mastery.

Attaining Coherence

There are a number of various exercises that one can use to gain coherence. One of the best I have come across is known as the **Quick Coherence HeartMath Technique**. Research studies have found that patterns of neurological signals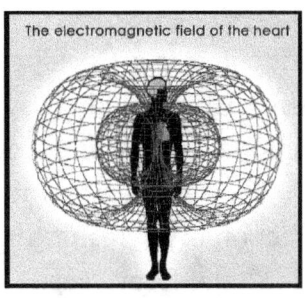

The electromagnetic field of the heart

change to more stable and ordered patterns when a person practices the HeartMath Quick Coherence Technique (Cardiac coherence, self-regulation, autonomic stability, and psychosocial well-being. Rollin McCraty and Maria A. Zayas. Sept 2014).

The exercise involves one shifting their attention to the canter of their chest (heart area) accompanied by conscious self-induction of a positive or calm emotional state. This causes a reinforcement of the association (pattern match) between an increased coherent rhythm and positive or calm emotions. When this takes place, positive feelings automatically initiate increased cardiac coherence. This increased coherence facilitates feelings of positive emotions and repeated practice causes a **re-patterning-type process** to take place. This is a key element in situations one has been a victim of trauma or sustained exposure to truly high-risk environments in the past. This re-patterning type process causes the system to allow people to maintain self-directed control and stability during their daily activities, **even in future challenging situations**. If this shift in the underlying baseline does not take place, it can be exceedingly difficult for a person to sustain positive behavioural changes, placing them at living their lives through the automatic filters of past familiar   experiences.

Technology for Developing Heart / Mind Coherence

When I first used these technologies I thought it was an old fashioned pager as it looked like someone was about to call me. However using the technology has really had a powerful effect on my self-mastery skills.

The HeartMath Institute has deployed several technologies that can help a person know when they are in coherence. Examples include Inner Balance® or emWave® Pro. Others include Relaxing Rhythms (Wild Divine) and the Stress Resilience Training System (Ease Interactive). These utilize a simple non-invasive earlobe or finger pulse sensor which display a user's heart rhythm which provides real-time feedback on the state of the person's coherence as they practice a simple exercise. From my experience over the years, I have found the emWave® to be of benefit, although I have yet to test the other systems.

Techniques for Enhancing Coherence

These next techniques I am about to show you are some of the best that will help one on the path towards self-mastery. They can be used in conjunction with the emWave®. They include the following:

The   Quick   Coherence   Technique

The Heart Lock-In Technique

The Freeze Frame Technique

The Inner-Ease Technique

These 4 separate techniques are non-invasive and made so they can intervene the moment a person begins experiencing unproductive thoughts, stress reactions or disturbing emotions. With simple practice, anyone can rapidly shift into a coherent physiological state before, during, or after a challenging or adverse situation. This in turn creates enhanced mental clarity, stability and a stable composure thus allowing one to exercise rapid emotional restraint, even in the midst of the most fearful situations. These are some of the ultimate self-mastery tools available on our planet today. Each coherence technique is suited to a particular situation. Let's next explore how to practice these techniques.

The Quick Coherence Technique
This technique helps a person to speak authentically while feeling fully heard. It also promotes discernment and greater rapport   and   empathy   between   others   (6).
     This technique involves placing one's attention in the area of their heart and then to imagine that their breath is flowing in and out of their chest area while they breathe a little deeper and slower than usual. The main objective of this technique is conscious regulation of one's respiration at a 10-second rhythm (for example - five seconds in and five

seconds out = 0.1 hertz). This 10-second rhythm increases one's cardiac coherence which starts the process of the autonomic nervous system shifting into a more coherent state of being [7] [8]. When one has conscious control over their breathing, they automatically increase the depth and slow the rate of their breathing. Hence physiological mechanisms take place which modulate efferent vagal activity, which in turn affect a person's heart rhythm.

If we were to look at this in greater detail, the newly modulated vagal activity increases vagal afferent nerve traffic which in turn increases coherence in the patterns of vagal afferent nerve traffic which finally influences neural systems that regulate sympathetic outflow allowing one to experience stronger emotional experiences via an oscillation of neural structures that govern their cognitive processes, which in turn enhances the ability of a person to think clearly and more sharply [7].

When this state of mind is attained and a person is using the emWave®, a 'beep' will sound when the person reaches coherence. This makes it an extremely valuable feedback tool for a person who wants to learn how to create strong and true coherence on demand. This shows that the body is operating at a peak state of coherence with all their physiological systems operating in strong harmony with each other.

Also if any of you practice relaxation techniques, one way I have discovered to greatly enhance relaxation is to instead of just imagining your hand or foot relaxing, instead place your awareness around your hand or foot and then visualize the stress and tension leaving and then imagine the hand or foot relaxing. This small tip greatly enhances the relaxation of muscles of the body. Also from practicing the HeartMath Quick Coherence Technique over the years, I have found that a special blend of essential oils that contain Limonene and Linalool greatly enhance and strengthen the ability to go into and maintain coherence. The formula is shown below –

An Essential oil Blend for Generating Sustained Coherence
Fill a bottle or cup with between 1.5 and 2oz of water. Next mix the following essential oils and place a few drops on a cotton swab. You can also add a few drops of Linseed Oil to enhance the effects.

7 drops of Bergamot
7 drops of Linalool
2 drops of Ylang Ylang II. Also called Ylang Ylang extra.
9 drops of Tea Tree
5 drops of Sweet Fennel
6 drops of Rose Germanium
6 drops of Lemon Myrtle

Let's go over a simple summary of the Quick Coherence Technique which is the most rapid way to create coherence on demand. It takes less than 2 minutes.
1 - **Heart Focus**. First focus attention on the region around your heart. This is the region in the centre of your chest. You can also place your hand over the centre region of your chest. This helps keeps your attention in the heart region.
2 - **Begin Heart Breathing**. Breathe deeply, yet normally. Feel as if your breath is going out through your heart area as you breathe out and as you inhale, feel as if your breath is flowing

in through your heart. Breathe a little deeper than normal and slowly and casually. Continue this until you find a natural inner rhythm that feels good to you.

3- **Heart Feeling**. While maintaining heart focus and heart breathing, activate positive feelings. This is done by recalling times when you felt feelings of appreciation or love. A quick tip to accomplish this is to remember a special place you had visited in the past, or the love you feel for a close friend, treasured pet or family member.

Heart-focused breathing is all about directing your attention to the heart area while breathing just a little extra deeply than   normal.

The Heart Lock-In Technique

This next self-regulation technique is more appropriate when one has more time and wants to maintain a stronger focus while in their coherent state. This technique enables one to "lock in" a positive feeling that is associated with their heart so that they can boost their clarity, energy, heighten peace in an environment or retrain their physiology to **sustain coherence for longer periods of time**. This makes it an especially valuable tool for athletes. As one consistently practices the Heart Lock-In technique, it will cause new reference patterns to take place which promote increased physiological efficiency, emotional stability mental acuity and a new baseline or norm. Use it to build up reserves of energy, deepen intuition, and stay longer in an intuitive flow.

1. Shift attention away from your mind and focus on your heart region. Breath slowly through the heart for ten to fifteen seconds.

2.   Next bring back vivid memories and feelings of care or love for someone whom you find easy to love or if you find this hard, focus feelings of appreciation on something or someone positive in your life. Run with that feeling(s) for between five and fifteen minutes.

3. Next send those feelings of care and love or appreciation to yourself or to others.

4. If you find logical thoughts returning, return focus gently back to the region around your heart and if the energy seems too intense or blocked, relax and imagine a softness in your heart.

5. After this exercise you can write down thoughts or intuitive feelings that are accompanied by a feelings of peace or inner knowingness. This will help you remember to act upon them.

The Freeze Frame Technique

The Freeze Frame Technique is an extension of HeartMath that powerfully reduces stress and anxiety. This technique is best used when we face a complex decision, feel emotionally charged or stressed, experience nervous anxiety or have trouble making an especially large decision. The technique works because, like movies, our conscious life takes place in a series of moving frames, which occur at specific time intervals and which are strung together over time. When we feel overwhelmed the Freeze Frame Technique 'freezes' the movie' or its frames of perception into just a single frame. The positive emotions experienced during this frozen moment in time bleed on over into future frames.

The Freeze Frame Technique

1. Temporarily disengage from any stressful feelings or thoughts.

2. Shift your awareness to the region around your heart.

3. Become aware of your breathing.

4. As you breathe, feel your breath coming in through your heart and out through the solar plexus. Practice this a few times until your body adapts to the new rhythm.

5. Next make sincere efforts to activate positive feelings. These may be genuine feelings of appreciation or care for someone or something in your life. It does not matter if you have trouble recalling positive times or experiences. This technique helps you even if you are feeling neutral.

6. Once this frame (positive feelings) has become frozen enough in your mind, place your hand on your heart and ask yourself the question, "**What would be an effective, efficient, emotion or action that would balance and de-stress my system?**" or "**What steps would help release stress and offer clarity about the decision I am trying to make?**"

7. Next allow yourself to quietly sense any change in your perception or feelings and sustain these new feelings as long as you can. Heart perceptions are subtle and suggest effective solutions in a gentle manner.

8: Next write out or mentally review any new insights you gained about your decision. Don't try too hard. Just see what flows naturally.

9: Compare how you feel about your new decision before and after you performed the Freeze-Frame Technique, perhaps even rating your feelings on a scale of 1 to 10. What new clarity was gained from the exercise?

A Second Quick Freeze Frame Technique

Step1 - Make sincere efforts to shift your mind away from racing thoughts and distractions or disturbed emotions in the region surrounding your heart.

Step 2 - As your body relaxes and calms down, experience the last time you generated positive feelings or lived a happy delightful experience. This could be a good or fun time you had in the past.

Step 3 - Next, using your common sense and intuition, sincerely ask your heart the following question. "What would be a better response to the situation? What response would repair the damage to _____?
"

Step 4 - Next listen to the answer generated in your heart region. If you are still anxious of what your heart is saying, repeat the process again. It sometimes takes a few repeats of the exercise before you gain a sense of clear intuitive knowing.

For major decisions that yield lots of emotional weight, immediate clarity is rare. Just repeat the Freeze-Frame Technique one step at a time and you will find that clarity will come with sincere practice. The answers received may not immediately be crystal clear, but you will definitely feel much less stressed, clearer and a lot calmer about your current situation.

The Inner-Ease Technique

Being in a state of natural ease is essential for bringing the essence of stillness into our daily life. Benefits and practical applications of Inner Ease include beginning your day by performing the Inner Ease Steps, along with an intention to breathe an attitude of ease when necessary throughout the day. While you proceed throughout the day, allow inner-ease to prevent impatience and distractions from overriding the true intuitive whispers that guide your higher choices and directions. Performing actions from inner ease is a heart-intelligent skill which creates a much more effective management and distribution of personal energy. Inner Ease reduces energy drain and regrets resulting from non-effective decisions. To put it simply, while you are in the flow of inner ease, you begin to travel the straightest path toward manifestation of your intentions.

Performing the Inner Ease Technique

1. While stressed, acknowledge that you feel stressed out. Honestly acknowledge feelings of impatience, anxiety, overload, frustration, anger, being judgmental, mentally gridlocked, etc.
2. Now take a pause and do Heart-Focused Breathing. Simply breathe slower than usual and pretend you are breathing through your chest area.
3. Next imagine with each breath you breathe in that you are drawing in feelings of inner-ease and are infusing your emotional and mental nature with balance and self-care from your heart.
4. As you feel your stressful feelings subside, affirm with heartfelt commitment that you intend to anchor and maintain that state of ease while engaging in your projects, challenges or daily activities.

Autogenic Training

A research study involving participants who partook

autogenic training classes, found increased HRV coherence and that their cardiac coherence showed strong correlations with EEG alpha brainwave activity. The researchers concluded that cardiac coherence was a general marker for the meditative state [9].

Prayer and HRV

A study involving five types of prayer on HRV found that all types of prayer caused increased cardiac coherence. Specifically it was the prayers of **heartfelt love** and **gratefulness** that caused much higher coherence levels [10]. Now that we have a better understanding of what coherence is, let's explore it in the next chapter on how we can use it to increase our intuition.

References. Chapter 8.

1. McCraty, R. and M. Atkinson, Resilence Training Program Reduces Physiological andPsychological Stress in Police Officers. Global Advances in Health and Medicine, 2012. 1(5): p.44-66.

2. McCraty, R. and M. Zayas, Cardiac coherence, self-regulation, autonomic stability, and psychosocial well-being. Frontiers in Psychology, 2014. 5(September): p. 1-13

3. McCraty, R. and D. Childre, Coherence: bridging personal, social, and global health. Altern Ther Health Med, 2010. 16(4): p. 10-24.

4. McCraty, R. and D. Tomasino, Coherence-building techniques and heart rhythm coherence feedback: New tools for stress reduction, disease prevention, and rehabilitation, in Clinical Psychology and Heart Disease, E. Molinari, A. Compare, and G. Parati, Editors. 2006, Springer-Verlag: Milan, Italy.

5. Childre, D. and H. Martin, The HeartMath Solution1999, San Francisco: Harper SanFrancisco.

6. Childre, D. and B. Cryer, From Chaos to Coherence: The Power to Change Performance2000, Boulder Creek, CA: Planetary.

7. McCraty, R., Atkinson, M., Tomasino, D., & Bradley, R. T, The coherent heart: Heart-brain interactions, psycho physiological coherence, and the emergence of system-wide order. Integral Review, 2009. 5(2): p. 10-115.

8. Alabdulgader, A., Coherence: A Novel Nonpharmacological Modality for Lowering Blood Pressure in Hypertensive Patients. Global Advances in Health and Medicne, 2012. 1(2): p. 54-62.

9. Reorganization of the brain and heart rhythm during autogenic meditation. Dae-Keun Kim. Jan 2014.

10. Stanley, R., Types of Prayer, Heart Rate Variability and Innate Healing Zygon 2009. 44(4).

Chapter 9. How to use Coherence to Enhance Intuition and Psychic Ability

Today when we watch television there is an approximate 5 to 15 minute delay. This is to cover up any mistakes made by the live reporter, actors or other live television personalities. What if we could reverse this process and instead have a reverse-delay of 5 minutes. In other words the television screen we watch would be showing what happens 5 minutes into the future and if we did not like it, we could avoid any mistakes we see! While this is not a reality that we know of today, there exists the next best thing, our intuition.

The Latin verb intueri means "intuition" which is translated as to contemplate or to 'look inside'. Reports from people who practice exercises that increase self-awareness to enhance the sensitively of their inner signals state that their heart appears to be communicating a steady stream of intuitive information to their mind and brain. Because there exists a relationship between heart coherence and intuitive signals [1] a person's capacity to shift into coherence is a key factor in accessing intuition. It is possible to access intuitive intelligence more efficiently by going into coherence, quieting emotional unrest and mental chatter and by paying attention to shifts in one's feelings. This in turn draws out intuitive signals to one's waking conscious awareness [2].

Research by the HeartMath Institute discovered that heart-rhythm coherence is connected with pre-stimulus-related afferent (ascending) signals that flow from the heart to the brain [3]. Hence these signals are important elements of intuition which are particularly salient in **pattern recognition**. They are also involved in all types of intuitive processing. They are the primary keys to harnessing intuition and allow one the ability to save time, energy and resources. When it comes to danger, our intuitive abilities operate at their peak. Only those who choose to ignore their intuition heeding them of future danger may end up paying the price. The ability for a person to heed their intuition is what some people refer to as their utilizing their higher capacities or higher self. Physicist David Bohm describes this as "our implicate order and undivided wholeness" [4].

Accomplished scientists propose that these higher functions operate in the frequency domain outside of space and time. They also suggest some of the possible mechanisms which govern how this method of functioning interacts with biological processes [5] [6] [7 to 11]. Studies are starting to look at if indeed the human heart consists of communication channels which connect it with the energetic heart (higher self) [12]. Hence, part of the wisdom which streams from our soul's higher information consists of non-local intuition. Therefore it is transformational and contains within it energy fields which filter down into the body's physiological system via the energetic heart. This in turn informs our moment-to-moment experiences and interactions (also called heart intelligence).

What is Heart Intelligence?

We all are familiar with the term "IQ", which is a measurement of one's ability to solve problems and identify familiar patterns in random data. However did you know our heart also has its own IQ? Just as our brains utilize aspects of the mind to make decisions and interpret familiar patterns, heart intelligence exists as a flow of higher awareness. The intuition one experiences is when the emotions and mind become aligned into synchronistic alignment with the energetic heart (higher self). It occurs mainly if one is coherent / heart-centered and has a tighter coupling and closer alignment with one's deeper source of intuitive intelligence. Hence, one is able to more intelligently self-regulate one's emotions and thoughts. Over time this lifts one's consciousness and creates new internal psychological and physiological baselines [12]. To put it simply, an increased flow of intuitive information is communicated via the emotional energetic system to the brain's systems which results in a clearer connection with our deeper inner voice.

Intuition Puts one on the Right Path

Intuition is helpful for eliminating unnecessary energy expenditures. Not heeding intuition can contribute to unnecessary energy wasters which can deplete valuable internal reserves, which in turn make it much more difficult to apply self-regulation techniques and achieve self-mastery. This in turn makes it harder for one to be in charge of their emotions, attitudes and behaviours which take place in ordinary day-to-day life situations. Heeding intuition leads to increases in one's ability to handle awkward situations such as dealing with difficult people with greater ease as it promotes harmonious interactions with others and clearer connectivity with others.

Gerard Hodgkinson of Leeds University in England states a growing body of research suggests underlying non-conscious aspects of intuition exist. Non-conscious aspects of intuition are forms of intuitive perception which are forms of implicit knowledge or implicit learning [13]. Intuitive perception plays important roles in entrepreneurship, learning, business decisions, medical diagnosis, healing, spiritual growth and general overall well-being [14] [15]. Research is also starting to discover that intuition plays important roles in creativity, decision-making and social cognition. Hodgkinson states "intuiting" exists as a complex set of interrelated cognitive, somatic and affective processes in which there is no apparent intrusion of deliberate rational thought taking place. Another definition is "affectively charged judgments that arise through rapid, holistic and non-conscious associations [16]." An ability to intuit may also be regarded as an inherited unlearned gift [17] [18]. Intuition and creativity are closely linked in that intuition acts as a beacon or guiding light from which new ideas and inspiration flow forth from different realities and dimensions.

The 3 Main Types of Intuition

Energetic Sensitivity

This type of intuition refers to the ability of the human nervous system to respond to and detect environmental signals such as electromagnetic fields (Energetic Communication). After performing an exercise I call the Emerald Tablets exercise, of which I wrote an entire book on titled The Emerald Tablets: The Keys of Life and Death by Thoth the Atlantean, I found that the restorative / rejuvenative energy would exhibit peaks whenever there were more cosmic rays. This may be due to less solar activity / radiation which lets in more solar radiation. Hence excessive solar activity is detrimental to human health and may be a major contributor to the aging process.

Studies conducted by Stoupel, et al., discovered high levels of cosmic rays combined with increased geomagnetic activity were accompanied by an increase the number of emergency room visits and death. His study also found increases in cerebral strokes and sudden cardiac deaths [19] [20]. Studies now confirm that in both animals and humans that nervous-system activity is affected by geomagnetic activity [21] with the largest effects taking place during above average geomagnetic activity.

Recent research has also discovered that changes in earth's magnetic field **are able to be detected** approximately one hour or in some cases longer **before** the presence of a large earthquake [22]. Hence, if the human heart exhibits pre-sentiment features, than we can say that **earth's magnetic field is earth's nervous system**. Energetic sensitivity can also take place when one feels they are being watched and a few studies have verified this type of sensitivity [23].

The Fundamental Basics of Non-local Intuition
The term non-local in the world of physics means wireless and without distance limitations and in some cases without time limitations. For example early telephones used a wire to connect two people. Today telephone communication is mostly wireless. Another example of non-local intuition is when a parent senses something is happening to their child who may be many miles away. Another example is the successful urging that is often experienced by successful entrepreneurs about factors that relate to making a major business   decision.

Communication in this manner takes place across the body's cells and neural networks in a non-local fashion, akin to a hologram.

To further explain the energetic communication in biological systems, we can examine Pribram's holographic approach. He states the communication of information that occurs in biological systems is due to information processing principles related to holographic theory (McCraty et al., 1998 Pribram, 1991 Pribram & Bradley, 1998).

Holograms are fun and exciting forms of art. Do we exist in a hologram that is part of a larger hologram representing the universe? The hypothesis that the human heart encodes and distributes information via a holographic model is based on the model described by neuropsychologist Karl Pribram who stated that the neural processes in our brains create our memories and drive our perception (Pribram, 1971, 1991). Pribram further states that neural impulses relay information from one section of the brain to other sections and that actual processing of information takes place in the spectral domain of energy frequency. This spectral domain is located **outside space and time** and that these waves of energy produced by the functioning of the neural microstructure interact with each other. Pribram further goes on to add that the mathematics that Gabor (1948) used to explain quantum-holographic principles show that the physics of signal processing describes information processing occurring as electromagnetic interactions between the axon and dendritic fields of the brain's neurons (McCraty et al., 1998).

As of 1995, Pribram, as well as other researchers have presented supporting experimental evidence that supports the veracity of Pribram's claims regarding the bio-energetic model of the brain's information processing system (King, Xie, Zheng, & Pribram, 1994 McCraty et al., 1998 Pribram,

1971, 1991 Santa Maria et al., 1995).

Pre-Sentient

Reserach suggests the heart's energy field is connected to fields of information not bound by space and time. This theory is based on a rigorous experimental study that sought to determine if the body received and processed future information **before an event occurred** (McCraty et al., 2004a, 2004b). The results of the study provided astounding and surprising results that showed that both the brain as well as the heart received and responded to information about a future event before it actually happened. The study also found evidence that **the heart receives intuitive information before the brain** (McCraty et al., 2004a, 2004b). This conclusively proves the heart is directly connected to subtle energetic fields filled with ambient information surrounding the human body and maybe even beyond. This energetic field interacts with the multiplicity of greater energy fields to which the human body is embedded in. This includes the quantum vacuum and possibly solar and galactic energy fields.

Energetic information retrieval has been proposed as a method of informational transfer from all psychosocial, physical and biological interactions which exist as enfolded sub-spaces in spectral order outside the time/space world in the energy waveforms of the quantum vacuum. The majority of these theories are based on Holographic principles (Gabor, 1948) and are one of just a few methods used to describe how information about an organization of a whole structure takes place in a non-localized fashion which is enfolded and distributed to equal regions and parts via energy waveforms that are produced by interactions in the brain, (Pribram, 1971, 1991) social structures, (Bradley & Pribram, 1998, Bradley, 1987) and which includes the universe (Nadeau & Kafatos, 1999, Bekenstein, 2003). Hence, energy waveforms that arise due to the heart's electromagnetic field, encode and distribute information in a holographic fashion throughout the body as a whole.

Meta-data   Pre-sentiment   Studies

A meta-analysis study involving nine experiments which looked at physiological responses that took place before a future event (pre-stimulus responses) occurred which was not otherwise anticipated through any known inferential process, found statistically significant results during eight of the nine experiments involving more than 1,000 participants [24]. Another researcher looking at 26 similar studies concluded that a strong pre-stimulus response took place in the body's physiological activity **before** unpredictable stimuli unfolded. This was despite the fact there existed any known explanation of the mechanisms for this finding [25].

The Human Heart as a Receiver for Future Information

Compelling evidence suggests the human heart is connected to fields of information that are not bound by the classical limits of space or time [26] [27]. This has been proven in another rigorous experimental study that was conducted proving that the heart processes and receives information about a future event before the event actually took place [26] [27]. Hence the human heart exhibits pre-stimulus responses. These responses carry information that work in tandem with or are transferred to the nervous system from the heart which the brain then decodes and receives information / images about the future. Further studies may reveal that pre-anticipatory responses are clearer and sharper during favorable solar windows during full moons, which most likely results in a slightly more stimulated parasympathetic nervous system.

A scientific study conducted by the HeartMath Institute measured a person's heart rate variability (HRV) and skin conductance levels during emotional responses. The response was designed to evoke an emotional reaction when a person saw either a calm or emotionally arousing picture. The study found significant differences took place between the person's HRV responses in the pre-stimulus period before future emotional or calm photos were shown. The physiological responses in the body took place approximately **4.8 seconds before** the participants actually saw the photographs. The study also found that not only did both the heart and brain exhibit pre-stimulus responses 4 to 5 seconds before the future emotional picture was randomly shown by computer, but that a **person's heart will receive the information** approximately 1.5 seconds **before the brain** receives it [28]. This is an important finding because it shows that emotions travel faster than logic or reasoning. Other independent studies have since come to the same conclusion in that the heart will elicit a response to a future or distant event [29] [30] [31] [32] [33] [34] [35].

It may also be that when cosmic rays are more abundant that the presentiment lead time is longer. Our research at the Solar Institute over the years has sought techniques to increase this pre-anticipatory time from seconds to days. The key element is favourable solar weather and lunar conditions combined with a tool we call the ARV Amplification Device. This allows the the fine-fibers of the human nervous system to become much more sensitive during an ARV session. Combined with HeartMath coherence exercises, it results in extraordinary successful Associative Remote Viewing Sessions.

Hypothesis

Gravitational and electromagnetic coupling that takes place in the geodynamic torque waves generated by earth's core rotation is creating matrixes of coiling spaces through which an energetic anomaly, a form of an Einstein Rosen bridge if you might call it, open up through higher dimensional curved space / time regions in our 3D reality. This allows one to access another location or point in space-time via a 5th dimensional mind link / bridge.

How Coherence Enhances Intuition

Other studies have found when a participant was in the physiological coherence mode before a test involving pre-stimulus activity, that their afferent input from their cardiovascular system and heart exhibited changes in their brain's electrical activity. This activity was especially pronounced in the frontal areas of their brain [36]. To put it simply, the person was much more attuned to information from the heart while the person was in a coherent state before participating in the experimental protocol. Hence being in a state of physiological coherence enhances one's intuitive ability. This also shows that our heart is directly coupled to a field of energetic information which interacts with the multiplicity of similar fields which the body is embedded in.

Also research conducted by the HeartMath Institute discovered that the magnitude of pre-stimulus responses to future events is strongly related to the **degree of emotionality associated with that event** [37]. This makes perfect sense, because during a full moon people's emotions seem more intense. Once again we see that emotions play a key role and that they may not be limited by space / time.

Entrepreneurs and Intuition

A HeartMath study involving 30 experienced Iranian entrepreneurs conducted in the technology and science parks in Tehran, extended and duplicated the first study of intuition discussed earlier [38]. Entrepreneurs were chosen because research has found that entrepreneurs have strong tendencies to rely on their intuition when making major business   decisions.

The study consisted of 2 groups. The first group consisted of single individuals (N = 15). The second group consisted of individuals paired with co-participants pairs (N = 30). This second group was formed to study the "amplification" of intuition effects that take place via social connection. In the single participant experiment, the participants watched pictures on a monitor by themselves. In the co-participant pair experiment, two people watched the same pictures simultaneously on two screens while they faced one another sitting down. Both experiments yielded significant pre-stimulus activity. As expected, significant separation between the calm HRV and emotional curves was observed in the single-participant experiment. However in the co-participant pair experiment, a much larger separation was observed with the difference between the two groups being significant. The study concluded that single-participant experiments involving pre-stimulus activity show electrophysiological changes with prominent changes taking place in the heart rhythm, which in turn demonstrates intuitive foreknowledge. Results regarding the co-participant pair showed new evidence that **amplification of non-local intuition signals** take place between **two individuals**.

In a similar experiment, the information from two subjects that were seated facing each another located 5 feet apart performed the Heart Lock-In Technique [39], which produces sustained states of physiological coherence [40]. The study found that the subjects were able to synchronize themselves to extremely weak external electromagnetic fields produced by another person's heart.

Summary
Dual independent systems that are in coherence exhibit enhanced sensitivity to external electromagnetic fields.

The Full-Moon Effect and its Amplification Effects on Intuition
Everybody wants to win every time they go to the casino. Wouldn't it be great if one could tap into their psychic gifts and win every time? While that may not be a reality yet, published studies have found that the moon seems to put the odds of winning at roulette in one's favour.

Pre-Stimuli and Moon Phase
A study conducted by the HeartMath Institute utilized a roulette protocol consisting of 13 participants involving eight

separate tests which included two pre-stimulus sections [41]. The study looked at moon phase and if it had any effect on pre-stimulus activity as well as if it increased winning at roulette. Half of the tests were performed during the new-moon phase and the other half during the full-moon phase. The study found significant pre-stimulus activity.

The pre-stimulus activity began approximately **18 seconds before** participants experienced the future results. A significant difference was found in both pre-stimulus periods **during full moons**, which was a period the participants won more money. However, the results were entirely different during the new moon. This is in tandem with our studies at the Solar Institute where our ARV accuracy significantly declines the last few days before the new moon. The roulette study also discovered that if the participants had became more attuned to their internal cardiac related pre-stimulus responses, that they would have had even greater success on their betting choices. One key point here is solar flare activity must be low with solar wind speeds in the 350 range during the time of the full moon if the experiment is to succeed as we show in greater detail throughout this book.

Summary
The effects of the full moon may be acting as an amplifier for Heart/Mind coherence lengthening the pre-stimulus response time. This amplification effect may be due to the moon's influence upon the heart.

The Full Moon and its Effects on Physical Endurance

Because above average geomagnetic activity is detrimental to ARV sessions and a full moon enhances the success of ARV session, and the heart is affected by geomagnetic activity, it may be that full moons increase endurance.

A research study titled: A Study on the Physical Fitness Index, Heart Rate and Blood Pressure in Different Phases of Lunar Month on Male Human Subjects that was conducted by U. Chakraborty and T. Ghosh, and which was published in September 2013 [42], found that the rate of recovery of the participant's heart rate after performing the Step Test was quicker during a full moon, compared to the moon's first and last quarters. The study also found that while the participant's were resting, that their systolic and mean arterial **blood pressures were lower during full moons**. This could also explain the enhanced anti-aging effects one takes after taking the Carnosine anti-aging effects during full moons. The Carnosine anti-aging mix can be found at www.ez3dbiz.com/pdf_docs/healing_herbal_formulations.pdf

What is the Step Test?

The Step Test is a Harvard cardiac stress test that is used for detecting cardiovascular diseases. It also acts as a measurement of a person's fitness and their ability to recover after strenuous exercise. In simple summary, the faster a person's heart rate returns to resting, the more fit the person is.

As we covered earlier, people who have negative emotions are more susceptible to illness or show increases in their symptoms [43] [43b]. The opposite is also true, people who exhibit positive emotions and self mastery during full moons may be more prone to positive outcomes, thus turning what may be a negative situation into a positive one, especially if a person's intuition gives them forewarning of a negative future event during a full moon.

Why Self-Awareness Enhances Intuition

Conscious awareness of anything, including intuitive promptings and emotions is not possible until something has captured our full attention [43]. Hence, the saying 'observation creates   reality'.

It is very interesting to note that creativity and intuition are closely linked with one another. Accessing more of our inner sense of knowing and intuitive intelligence can be achieved by developing deeper levels of self-awareness of our perceptions, especially our more subtle feelings. These are feelings that don't normally rise to our waking conscious awareness. They are most easily accessed via one's intuition.

Now that you have a much better understanding of the relationship between coherence and intuition, the next chapter will show in greater detail the effects of coherence on the body's physiological systems. If you find it too over-technical, you may want to skip forward to Chapter 11 and discover how our earth's Schuman resonance is overlapping with our brainwaves.

References. Chapter 9.

1. McCraty, R., M. Atkinson, and R.T. Bradley, Electrophysiological evidence of intuition: Part 2. A system-wide process? Journal of Alternative and Complementary Medicine, 2004. 10(2): p.325-336.

2. Petitmengin-Peugeot, C., The Intuitive Experience, in The View from Within. First-person approaches to the study of consciousness, F.J. Varela and J. Shear, Editors. 199, Imprint Academic: London,. p. 43-77.

3. McCraty, R., M. Atkinson, and R.T. Bradley, Electrophysiological evidence of intuition: Part 2. A system-wide process? Journal of Alternative and Complementary Medicine, 2004. 10(2): p.325-336.

4. Bohm, D. and B.J. Hiley, The Undivided Universe1993, London: Routledge.

5. Pribram, K.H., Brain and Perception: Holonomy and Structure in Figural Processing. 1991,Hillsdale, NJ: Lawrence Erlbaum Associates, Publishers.

6. Laszlo, E., Quantum Shift in the Global Brain: how the new scientific reality can change us and our world2008, Rochester, VT: Inner Traditions.

7. Mitchell, E., Quantum holography: a basis for the interface between mind and matter, in Bioelectromagnetic Medicine, P.G. Rosch and M.S. Markov, Editors. 2004, Dekker: New York, NY. p. 153-158.

8. Tiller, W.A., J. W E Dibble, and M.J. Kohane, Conscious Acts of Creation: The Emergence of a New Physics 2001, Walnut Creek, CA: Pavior Publishing. (pp. 201-202).

9. Bradley, R.T., Psycholphysiology of Intution: A quantum holgraphic theory on nonlocal communication. World

Futures: The Journal of General Evolution, 2007. 63(2): p. 61-97. 257. Marcer, P. and W. Schempp, The brain as a conscious system. International Journal of General Systems, 1998. 27: p. 231-248.

10. Pribram, K.H. and R.T. Bradley, The brain, the me and the I, in Self-Awareness: Its Nature and Development, M. Ferrari and R. Sternberg, Editors. 1998, The Guilford Press: New York. p. 273-307.

11. Schempp, W., Quantum holography and neurocomputer architectures. Journal of Mathematical Imaging and vision, 1992. 2: p. 109-164.

12. McCraty, R. M. Atkinson, and R.T. Bradley, Electrophysiological evidence of intuition: Part 2.A system-wide process? Journal of Alternative and Complementary Medicine, 2004. 10(2): p.325-336.

13. Hodgkinson, G.P., J. Langan-Fox, and E. Sadler-Smith, Intuition: A fundamental bridging construct in the behavioural sciences. British Journal of Psychology, 2008. 99(1): p. 1-27.

14. Myers, D.G., Intuition: Its Powers and Perils2002, New Haven: Yale University Press.

15. Bradley, R.T., et al., Nonlocal Intuition in Entrepreneurs and Nonentrepreneurs: Results of Two Experiments Using Electrophysiological Measures. International Journal of Entrepreneurship and Small Business, 2011. 12(3): p. 343-372.

16. Dane, E. and M.G. Pratt, Exploring intuition and its role in managerial decision making. Academy of Management Review, 2007. 32: p. 33–54.

17. Bastick, T., Intuition: How we think and act1982, New York: Wiley.

18. Moir, A. and D. Jessel, Brainsex: The real difference

between men and women. 1989, London:: Mandarin Paperbacks.

19. McCraty, R. and M. Atkinson, Resilence Training Program Reduces Physiological and Psychological Stress in Police Officers. Global Advances in Health and Medicne, 2012. 1(5): p. 44-66.

20. Luthar, S.S., D. Cicchetti, and B. Becker, The construct of resilience: a critical evaluation and guidelines for future work. Child Dev, 2000. 71(3): p. 543-62.

21. Halberg, F., et al., Time Structures (Chronomes) of the Blood Circulation, Populations' Health, Human Affairs and Space Weather. World Heart Journal, 2011. 3(1): p. 1-40.

22. Craig, J. and N. Lindsay, Quantifying "gut feeling" in the opportunity recognition process. Frontiers of Entrepreneurship Research, 2001: p. 124-135.

23. Wiseman, R. and M. Schlitz, Experimenter effects and the remote detection of staring. Journal of Parapsychology, 1997. 61: p. 197-207.

24. Bem, D.J., Feeling the future: Experimental evidence for anomalous retroactive influences on cognition and affect. J Pers Soc Psychol, 2011.

25. Mossbridge, J., P. Tressoldi, E, and J. Utts Predictive Physiological Anticipation Preceding Seemingly Unpredictable Stimuli: A Meta-Analysis. Frontiers in Psychology, 2012. 3:390.

26. McCraty, R., M. Atkinson, and R.T. Bradley, Electrophysiological evidence of intuition: Part 1. The surprising role of the heart. Journal of Alternative and Complementary Medicine, 2004. 10 (1): p. 133-143.

27. McCraty, R., M. Atkinson, and R.T. Bradley, Electrophysiological evidence of intuition: Part 2. A system-

wide process? Journal of Alternative and Complementary Medicine, 2004. 10(2): p. 325-336.

28. McCraty, R., M. Atkinson, and R.T. Bradley, Electrophysiological evidence of intuition: Part 2. A system-wide process? Journal of Alternative and Complementary Medicine, 2004. 10(2): p.325-336.

29. Tressoldi, P.E., et al., Heart rate differences between targets and non targets in intuition tasks. Fiziol Cheloveka, 2005. 31(6): p. 32-6.

30. Hu, H. and M. Wu, New Nonlocal Biological Effect. Neuro Quantology 2012. 10(3): p. 462-467.

31. Tressoldi, P.E., et al., Implicit Intuition: How Heart Rate can Contribute to Prediction of Future Events. Journal of the Society for Psychical research. 2009. 73: p. 1-16.

32. Sartori, L., et al., Physiological correlates of ESP: heart rate differences between targets and nontargets. Journal of Parapsychology, 2004. 68(2): p. 351.

33. Tressoldi, P.E., et al., Further evidence of the possibility of exploiting anticipatory physiological signals to assist implicit intuition of random events. Journal of Scientific Exploration, 2010. 24(3): p. 411.

34. Bradley, R.T., R. McCraty, M. Atkinson, & M. Gillin. Nonlocal Intuition in Entrepreneurs and Nonentrepreneurs: An Experimental Comparison Using Electrophysiological Measures. in Regional Frontiers of Entrepreneurship Research. 2008. Hawthorne, Australia.

35. Toroghi, S.R., et al., Nonlocal Intuition: Replication and **Paired Subjects Enhancement Effects**. Global Advances in Health and Medicne, 2014.

36. McCraty, R., M. Atkinson, and R.T. Bradley, Electrophysiological evidence of intuition: Part 2. A system-

wide process? Journal of Alternative and Complementary Medicine, 2004. 10(2): p. 325-336.

37. McCraty, R., M. Atkinson, and R.T. Bradley, Electrophysiological evidence of intuition: Part 1. The surprising role of the heart. Journal of Alternative and Complementary Medicine, 2004. 10 (1): p. 133-143.

38. Craig, J. and N. Lindsay, Quantifying "gut feeling" in the opportunity recognition process. Frontiers of Entrepreneurship Research, 2001: p. 124-135.

39. Childre, D. and H. Martin, The HeartMath Solution1999, San Francisco: Harper. SanFrancisco.

40. McCraty, R., et al., The impact of a new emotional self-management program on stress, emotions, heart rate variability, DHEA and cortisol. Integr Physiol Behav Sci, 1998. 33(2): p. 151-70.

41. McCraty, R., Electrophysiology of Intuition: Pre-stimulus Responses in Group and Individual Participants Using a Roulette Paradigm. Global Advances in Health and Medicne, 2014. 3(2): p. 16-27.

42. A study on the physical fitness index, heart rate and blood pressure in different phases of lunar month on male human subjects. Chakraborty U and Ghosh T.. Sept 2013.

43. Brotman, D.J., S.H. Golden, and I.S. Wittstein, The cardiovascular toll of stress. Lancet, 2007. 370(9592): p. 1089-100.

43b. Marchand, A. and P. Durand, Psychological distress, depression, and burnout: similar contribution of the job demand control and job demand-control-support models? J Occup Environ Med, 2011. 53(2): p. 185-9.

Chapter 10. Coherence within the Body's Internal Functions

Techniques   for   Expanding   Coherence
There exist numerous rhythmic-breathing exercises that induce coherence for brief periods of time. However by utilizing methods such as the Freeze Frame Technique shown earlier, a person can achieve extended periods of physiological coherence or in the case of the Heart Lock-In Technique, build up significant reserves of energy for future use.

Of all the organs in the body, the heart has the most extensive neural connection of any organ. Rodolfo Llinas, Chief of Physiology and Neuroscience at the New York University School of Medicine, discovered that areas of a person's cortex emit steady oscillations occurring at a frequency of approximately 40 cycles per second (40 Hz) (Ratey, 2001). His research discovered that remote regions of the cortex were phase-locked into a 40-Hz frequency. This means that these waves oscillate in sync (a coherent fashion). Llinas suggested that the brain's neurons perform in concert with each other because they follow a kind of conductor. They key point here is that the heart and brain are in coherence with each other and by using special techniques, this coherence can be amplified to such an extent that the major organs of the body can be affected in a beneficial manner. This leads to better health, more energy and better control over one's emotions.

The Thalamus, Coherence and Consciousness

How lucky some people are to lead a life of limited stress, and be able to bring the stress experienced immediately under control. Buddhist monks must be the envy of many an overstressed   individual!

Coherence in Meditating Monks

A recent study of Buddhist practitioners who were expert meditators found increases occurring in their gamma band oscillation (40 Hz) and long-distance phase synchrony when they generated conscious states of "**unconditional loving-kindness and compassion,**" (Lutz, Greischar, Rawlings, Ricard, & Davidson, 2004). This is an important validation for our research as we have found that listening to 40 Hz gamma a few hours before an ARV session enhances intuition during the   ARV   session.

The main elements that act as the brain's internal conductor are the many intralaminar nuclei, which are located in the brain's thalamus. These nuclei project and receive information to numerous regions of the brain. They absorb information, reply to it, and also monitor responses to their replies. This acts as an elaborate feedback loop in through which resonant cavity activity at ~40 Hz becomes modified by incoming sensory signals. If the brain's intralaminar nuclei become damaged, a person enters an irreversible coma. It is only when the "conductor", synchronizes the brain's activity in concert that waking conscious activity takes place. This results in a large enough number of neural networks to oscillate in an ordered globally coherent fashion. This then causes more networks to join them creating a strong, coherent form of waking consciousness (Ratey, 2001). The brain's thalamus plays an active role in the generation of these rhythms. When a network becomes too loosely coupled or too excessively coupled, the system finds it harder to recruit the regional neural support systems it needs to respond to the increased demand. If one wants to optimize their performance, one should avoid being too overly relaxed, which results in an increased coupling, or avoid being too overly stimulated / stressed, which results in a decreased coupling. The key is to maintain a balance between both.

Summary
Neural communication pathways that interact with the heart and brain cause HRV activity.

One of the more interesting features of favorable solar weather conditions is an enhanced Schuman resonance. Let's explore this next.

References. Chapter 10.

1. Childre, D. and H. Martin, The HeartMath Solution1999, San Francisco: Harper San Francisco.
2. Childre, D. and D. Rozman, Transforming Stress: The HeartMath Solution to Relieving Worry, Fatigue, and Tension. 2005, Oakland, CA: New Harbinger Publications.

Chapter 11. The Schuman Resonance and its Effects upon the Human Body

Around the year 2012, I wrote a book titled Solar Flares and their Effects upon Human Behaviour and Health. The research published since that time has now confirmed that above average solar weather negatively impact the human body.

In this next section of the book we shall look at scientific studies confirming that indeed above average solar activity can be harmful to the health of some individuals and in other cases favorable solar weather not only enhances intuition, but when combined with HeartMath coherence techniques may accelerate the rate of healing and the recovery from some types of diseases and illness, especially when the Schuman resonance is stronger.

Schuman resonance and Human Cortical Activity

A study looked at the manifestation of the Schumann resonance and its influence within the electro-cortical activity of individuals. The study reviewed an earlier study which showed enhanced synchrony between the Schumann resonance and human cortical activity when it was measured in Sudbury, Canada and in Cumiana, Italy.

The study found the Schumann resonance was more defined for individuals who had stronger Schumann resonances intensities, as defined by a test assessing their z-score. One group was described as "flat", and the other 'high'. The high group exhibited valleys and peaks with characteristic standing waves of 7.18 Hz, 13.52 Hz and 20.29 Hz respectively.

Reference

Similar Spectral Power Densities Within the Schumann Resonance and a Large Population of Quantitative Electroencephalographic Profiles: Supportive Evidence for Koenig and Pobachenko. Kevin S. Saroka. Et al. Jan 2016.

The most likely mechanism that explains how solar and geomagnetic influences affect human health and behaviour are a coupling between the human nervous system and resonating geomagnetic frequencies, called Schumann resonances, which occur in the earth-ionosphere resonant cavity and its associated Alfvén waves. It is well established that these resonant frequencies directly overlap with those of the human brain and cardiovascular system. They are also known as Alfven waves and are associated with similar very low-frequency resonances. Numerous research studies now confirm that these frequencies directly overlap with the frequencies of the human brain (Science of the Heart. Page 82. Rollin McCraty. Feb 2016). These frequencies also overlap with activity that occurs in the body's autonomic nervous systems as well as cardiovascular activity (Science of the Heart. Page 82. Rollin McCraty. Feb 2016).

Changes in the Schuman Resonance and its Effect on Brainwaves

A study conducted by Pobachenko et al [1] examined earth's Schumann resonance (SR) and a person's EEG in a frequency range of between 6 and 16 hertz simultaneously. During a 24 hour period, the individuals observed showed variations in their EEG's which were similar to changes in earth's Schuman resonance. Hence, the biological EEG rhythm of an individual is characteristic of a daily rhythm of earth's Schuman resonance [1]. This is a major finding because it shows that the **Schuman resonance affects human brainwaves**. Hence, a more in-coherent Schuman resonance, which is usually caused by major disturbances in earth's geomagnetic activity would affect a person's brainwaves, which could explain the numerous studies showing that above average or below average geomagnetic activity negatively impacts human health (Solar Flares and Their Effect Upon Human Behavior and Health. Scott Rauvers).

The 7 Main Frequencies of Earth's Schuman Resonance
The main frequency is 7.83 hertz (with a daily variation of approximately ± 0.5 hertz). The other frequencies are ~ 14, 20, 26, 33, 39 and 45 hertz. These frequencies closely overlap with human brainwaves.

Human   Brainwave   Frequencies

8 to 12 hertz (alpha)

12 to 30 hertz (beta)

30 to 100 hertz (gamma)

Because there is not an exact 'match' of the Schuman resonance and human brainwaves, there may exist a 'cavity' type effect due to an individual's surroundings such as mountains, air pressure and other environmental variables that create this offset of the Schuman resonance. As discussed earlier, the brain is weaker than the heart. Hence the brain is more sensitive to changes in geomagnetic activity. Hence Schuman resonance variations alter neuro-hormone responses and brain-waves.

Tapping into a favourable Schumann resonance, which occurs when solar weather conditions are favorable, is obtained by either using coherence techniques or by being in or generating an environment where a favorable difference in pressure waves exist. For example piezoelectricity can be obtained from applying pressure such as the squeezing of a quartz crystal. When earth's barometric air pressure has peaked and is beginning its descent, a change in air pressure also takes place. This period of peaking air pressure is a time our ARV sessions always display excellent results (air pressure going from a peak towards a low). This may also explain why animals such as snakes can sense an approaching earthquake because earthquakes generate large amounts of piezoelectric currents and tsunamis create a change in the localized region of air pressure before they approach shore. This change in pressure may be how Dr. Yoshiro Nakamats, who has more patents than Thomas Edison, gets his ideas for great inventions.

Dr. Yoshiro Nakamats states his greatest ideas occur while taking long underwater swims. He says this is because if the brain has too much oxygen in it, it acts as a deterrent for true inspiration. The trick to starving the brain of oxygen for a short period of time is to dive deep underwater. This will then allow the water pressure to fill the brain with blood. He states that after diving to the bottom of a swimming pool, he holds his breath as long as possible. He calls this the "0.5 seconds before death" zone. While in "the zone" he will visualize an invention and then immediately write the ideas and thoughts he has received on a plexi-glass waterproof tablet. Next he returns to the surface of the swimming pool. The region at the bottom of a swimming pool is a region of natural high pressure.

References. Chapter 11.

1. Pobachenko, S.V., Kolesnik, A. G., Borodin, A. S., Kalyuzhin, V. V. The Contingency of Parameters of Human Encephalograms and Schumann Resonance Electromagnetic Fields Revealed in Monitoring Studies. Complex Systems Biophysics, 2006. 51(3): p. 480-483.

Chapter 12. Anticipatory Reactions and the Human Nervous System

Another word for pre-stimulis is 'anticipatory reaction'. Research studies have confirmed an "anticipatory reaction effect" (a change in activity before a future event) that occurs in the physiological pathways of the human body. A research study found that changes occur in the human body 2 to 3 days before a geomagnetic storm. Major changes include: the person's heart rate, their skin conductance, HRV parameters, blood pressure and physiological complaints [1] [2] [3 to 8]. This is a major discovery / finding, because it shows that the body is exhibiting pre-stimulis responses a few days **BEFORE** the geomagnetic storm. Also the time range matches that of how far out we are able to see during an ARV session, which is a maximum of 4 days.

Our heart receives limited signals of future events on its own. For clearer long-term signals, it takes cross-coherence of multiple systems oscillating in unison, a symphony if you will, before it is able to receive future information. It is our hearts that are the key mechanism from which we receive future information about events that are about to take place in the near future. This is done by our heart communicating with our body and brain via four major pathways:

1- Nervous system
2- Hormones
3 - Pressure / pulses waves
4 - Electromagnetic fields

When we are in coherence, all 4 of the above work together.

It has already been proven that our heart's magnetic field is able to be detected by the nervous systems of nearby people or animals [9].

Anticipatory Reactions

It takes a total of approximately 8 minutes for the radiation produced by coronal ejections from our sun to reach us here on our earth. It takes several days for the dense particles emitted by the solar wind to reach Earth's magnetosphere, causing geomagnetic storms. This time-delay itself may contain anticipatory reactions. Hence, earth may be reacting in some way to the oncoming particles before they strike earth's magnetosphere. It may be that as these higher energy particles fade out, the body's nervous system becomes more sensitive to future events. It is our hypothesis at the Solar Institute that the human nervous system can tap into an unseen field or force during the period after major solar activity as geomagnetic activity starts entering its quiet period. This is why our ARV sessions are always planned to take place a few days just after a geomagnetic storm because the chance of another geomagnetic storm occurring (cause of interference) is greatly reduced. This leads us to conclude that the "noise" generated by above average geomagnetic activity is affecting the human nervous system and that a calmer / more quiet / noise free nervous system is highly conductive to successful ARV sessions.

Studies have already proven that a chaotic nervous system can be detrimental to healthy physiological functioning and the utilization of one's energy. This chaos can result in chronic stress or age-related illnesses [10] [11] [12].

Anticipatory effects taking place in the human body were

first recorded by Chizhevsky during the 1920's before any knowledge of high frequency emissions such as gigahertz frequencies (the sun's 10.7cm solar radio flux) and X-rays radiated by our sun were known to exist. Chizhevsky believed there existed a form of unknown radiation that was being emitted by our sun and that it was responsible for causing the anticipatory reactions he observed [13]. This is a very interesting finding, because we have found that when the sun is more active, active being not more sunspots, but an increase in the sun's solar 10.7cm radio flux, that our ARV sessions are more accurate. This leads us to conclude that this specific region / frequency of solar activity may be enhancing the speed or capacity at which the human nervous system receives information from a future event or that the brain is able to process information.

The Sun's 10.7cm Solar Radio Flux and Fatigue

Studies have found that increased levels of the sun's 10.7cm radio flux are commonly associated with lower fatigue, increased positive moods and enhanced mental clarity. The same study found increased solar wind speeds (above 370) exhibited the opposite effects [14]. Our most successful ARV sessions would always occur when the sun's solar wind speed was between 330 and 350 and the 10.7cm radio flux was increasing, especially if it had increased for 3 or more consecutive days.

Causes of Reduced HRV

Increased heart rate activity (HR), and a reduction in heart rate variability (HRV) and an increased number of arrhythmic events has been shown to take place during magnetic storms [15]. And as discussed earlier, changes in the human body take place 2 to 3 days before a geomagnetic storm. The study found that the major changes included: the person's heart rate, their skin conductance, HRV parameters, blood pressure and physiological complaints [16] [17] [18] [19] [20] [21] [22].

Prime Physiological Conditions for an ARV session -

1- A healthy stimulated Parasympathetic Nervous System

2 - Increased Resting Heart Rate Variability (HRV)

3 - Besides diet and the proper mental preparation before an ARV session, performing HeartMath self-regulation techniques and using specific substances such as limonene, will allow the above two parameters to effortlessly take place.

A Solar Storm's Effects upon the Human Parasympathetic Nervous System. An in-depth study.

A research study that lasted 5 months looked at the time lags in the human body's autonomic nervous system and its responses to changes in magnetic and solar variables [23]. The study found that HRV was correlated positively with the sun's 10.7 cm solar radio flux. The changes observed were an **increase in parasympathetic activity**, with the main effects occurring approximately **20 hours after** the sun's F10.7cm radio flux had increased. This is a major discovery because it shows that HRV / parasympathetic nervous system activity is affected by the sun's 10.7cm radio flux and that the activity exhibits a lasting resonance in the body's parasympathetic nervous system. Hence when an ARV session is performed, one may be tapping into this reservoir of energy. This suggests again, as covered earlier, that the right stimulation the parasympathetic nervous system plays a key role in successful ARV sessions.

Summary

Strong evidence now exists human behaviour and health are globally influenced by geomagnetic and solar activity. The evidence is strong and convincing. Hence earth's magnetic fields are acting as carriers of biologically relevant information that are interconnected with all living systems. This may be where the consciousness of plants or even mammals resides. Hence, where does the information come from to repair a plant leaf when it is injured or damaged? Some say DNA. DNA also encodes and carries information [24].

Now let's dive a little deeper into the mechanisms of the nervous system as we are beginning to see that it appears to play a key role in remote viewing.

References. Chapter 12.

1. McCraty, R., M. Atkinson, and R.T. Bradley, Electrophysiological evidence of intuition: Part 2. A system-wide process? J. Altern Complement Med, 2004. 10(2): p. 325-36.

2. Goldstein, D.S., Stress, allostatic load, catecholamines, and other neurotransmitters in neurodegenerative diseases. Endocr Regul, 2011. 45(2): p. 91-8

3. Fredrickson, B.L., The role of positive emotions in positive psychology. The broaden-and-build theory of positive emotions. American Psychologist, 2001. 56(3): p. 218-226.

4. Fredrickson, B.L. and T. Joiner, Positive emotions trigger upward spirals toward emotional well-being. Psychological Science, 2002. 13(2): p. 172-175.

5. Fredrickson, B.L., et al., What good are positive emotions in crises? A prospective study of resilience and emotions following the terrorist attacks on the United States on September 11th, 2001. Journal of Personality and Social Psychology, 2003. 84(2): p. 365-376.

6. McCraty, R. and D. Tomasino, Emotional stress, positive emotions, and physiological coherence, in Stress in Health and Disease, B.B. Arnetz and R. Ekman, Editors. 2006, Wiley VCH: Weinheim, Germany. p. 342-365.

7. McCraty, R., et al., The effects of emotions on short-term power spectrum analysis of heart rate variability. Am J Cardiol, 1995. 76(14): p. 1089-93.

8. Rein, G., M. Atkinson, and R. McCraty, The physiological and psychological effects of compassion and anger. Journal of Advancement in Medicine, 1995. 8(2): p. 87-105.

9. McCraty R. The energetic heart: bio-electromagnetic

communication within and between people. In: Rosch PJ, Markov MS, editors. Bioelectromagnetic Medicine. New York: Marcel Dekker (2004). p. 541–62).

10. Singer, D.H., et al., Low heart rate variability and sudden cardiac death. Journal of Electrocardiology, 1988 (Supplemental issue): p. S46-S55.

11. Thayer, J.F., et al., Heart rate variability, prefrontal neural function, and cognitive performance: the neurovisceral integration perspective on self-regulation, adaptation, and health. Ann Behav Med, 2009. 37(2): p. 141-53.

12. Camm, A.J., et al., Heart rate variability standards of measurement, physiological interpretation, and clinical use. Task Force of the European Society of Cardiology and the North American Society of Pacing and Electrophysiology. Circulation, 1996. 93(5): p. 1043-1065.

13. Fredrickson, B.L. and T. Joiner, Positive emotions trigger upward spirals toward emotional well-being. Psychological Science, 2002. 13(2): p. 172-175.

14. Carroll, D., et al., Blood pressure reactions to the cold pressor test and the prediction of ischaemic heart disease: data from the Caerphilly Study. Journal of Epidemiology and Community Health, 1998. 52: p. 528-529.

15. Armour, J.A., Potential clinical relevance of the 'little brain' on the mammalian heart. Exp Physiol, 2008. 93(2): p. 165-76.

16. McCraty, R., M. Atkinson, and R.T. Bradley, Electrophysiological evidence of intuition: Part 2. A system-wide process? J Altern Complement Med, 2004. 10(2): p. 325-36.

17. Goldstein, D.S., Stress, allostatic load, catecholamines, and other neurotransmitters in neurodegenerative diseases.

Endocr Regul, 2011. 45(2): p. 91-8.

18. Fredrickson, B.L., The role of positive emotions in positive psychology. The broaden-and-build theory of positive emotions. American Psychologist, 2001. 56(3): p. 218-226.

19. Fredrickson, B.L. and T. Joiner, Positive emotions trigger upward spirals toward emotional well-being. Psychological Science, 2002. 13(2): p. 172-175.

20. Fredrickson, B.L., et al., What good are positive emotions in crises? A prospective study of resilience and emotions following the terrorist attacks on the United States on September 11th, 2001. Journal of Personality and Social Psychology, 2003. 84(2): p. 365-376.

21. McCraty, R. and D. Tomasino, Emotional stress, positive emotions, and psychophysiological coherence, in Stress in Health and Disease, B.B. Arnetz and R. Ekman, Editors. 2006, WileyVCH: Weinheim, Germany. p. 342-365.22. McCraty, R., et al., The effects of emotions on short-term power spectrum analysis of heart rate variability. Am J Cardiol, 1995. 76(14): p. 1089-93.

22. Rein, G., M. Atkinson, and R. McCraty, The physiological and psychological effects of compassion and anger. Journal of Advancement in Medicine, 1995. 8(2): p. 87-105.

23. Carroll, D., et al., Blood pressure reactions to the cold pressor test and the prediction of ischaemic heart disease: data from the Caerphilly Study. Journal of Epidemiology and Community Health, 1998. 52: p. 528-529.

24. Berg JM, Tymoczko JL, Stryer L.New York: W H Freeman 2002.. Chapter 5DNA, RNA, and the Flow of Genetic Information. Biochemistry. 5th edition.

Chapter 13. The Human Nervous System, Solar Weather and Associative Remote Viewing Sessions

The most likely mechanism that explains how solar and geomagnetic influences affect human health and behaviour are a coupling between the human nervous system and resonating geomagnetic frequencies, called Schumann resonances, which occur in the earth-ionosphere resonant cavity and its associated Alfvén waves. As covered earlier it is now well established that these resonant frequencies directly overlap with those of the human brain and cardiovascular system.

Coherence is associated with increased parasympathetic activity and is part of the relaxation response. Increased parasympathetic activity tends to take place during periods of relaxation and rest and even with structured meditation techniques. This typically causes an overall reduction in ANS (Autonomic Nervous System) outflow. Also meditation and relaxation do not always cause significant increases in coherence. This is because fundamental differences exist between coherence and the physiological correlates of relaxation. Also. associated psychological states during these times can be markedly different. Increased parasympathetic activity causes large peaks to take place in the HF band of the power spectrum.

Studies have found that increases in parasympathetic activity (**vagal tone**) [1] were associated with reduced cortisol levels and increases in DHEA [2]. Other studies found decreases in one's blood pressure and stress measures in people who exhibited hypertensive behaviours [3] [4].

Studies have found that the positive effects of increased

parasympathetic activity, when necessary, result in reduced health-care costs [5] as well as major improvements in people suffering from congestive heart failure [6].

Further Reading

Synchronization of Human Autonomic Nervous System Rhythms with Geomagnetic Activity in Human Subjects. Rollin McCraty et al. July 2017.

Parasympathetic fibers that make up the parasympathetic nervous system are primarily found in the vagus nerves, which are situated just under the ear lobe. Parasympathetic fibers that regulate subdiaphragmatic organs travel throughout the body's spinal cord and research is looking at if stimulation of the vagus nerve may help treat people who have epilepsy. Acetylcholine is released by the body's parasympathetic nervous system and high levels of acetylcholine in the body causes changes the brain's theta oscillations thus enhancing memory.

The Vagus Nerve

Contrary to what it sounds like, it does not mean "Vegas Nerves". The longest nerve in the human body is the vagus nerve. It is the main component of the body's parasympathetic nervous system, which together with the body's sympathetic nervous system, make up the human autonomic nervous system.

A research study that was undertaken in 2010 found that a participants practicing loving kindness meditation training experienced higher levels of positive emotions which affected their vagal regulation (Kok et al., 2013). Practicing Tai Chi has also been shown to increase vagal activity (Tai Chi research review. Field T. August 2011).

Another study found that people who practiced yoga regularly had increased HRV and an enhanced vagal tone compared to non-yoga practitioners [5c]. Another study found that **loving kindness meditation increased the participant's vagal tone**, due to the increase in positive emotions (Kok et al., 2013). Also increased feelings of social connectivity have been shown to increase vagal tone [5d]. The vagus nerve releases acetylcholine [5e] and the herb Butterfly Pea has been shown to enhance acetylcholine levels in studies done on Mice [5f]. I can personally attest to this as I have had butterfly tea and it is of a deep violet colour. After taking it, my entire body felt calm for days afterwards. It also contains anthocyanins, which are a powerful anti-aging substance. Butterfly Pea has also been shown to exert similar effects to the cerebro protective drug Pyritinol [5g].

Moderate Pressure Massage stimulates the parasympathetic Nervous System

A research study looked at 29 adults that were randomly given light pressure or moderate pressure massage therapy for 15 minutes. The study found that the volunteers who received moderate pressure massage had a more stimulated parasympathetic nervous system characterized by an increase in HF. This suggested increased vagal efferent activity as well as a decrease in the LF/HF ratio. Hence this caused a shift from sympathetic to parasympathetic activity which peaked approximately 7 minutes into the massage. The volunteers that received the light pressure massage showed a sympathetic nervous system that was characterized by decreased HF and increased LF/HF [5h].

References. Chapter 13.

1. Tiller, W.A., R. McCraty, and M. Atkinson, Cardiac coherence: a new, non-invasive measure of autonomic nervous system order. Altern Ther Health Med, 1996. 2(1): p. 52-65.

2. McCraty, R., et al., The impact of a new emotional self-management program on stress, emotions, heart rate variability, DHEA and cortisol. Integr Physiol Behav Sci, 1998. 33(2): p. 151 -70.

3. Alabdulgader, A., Coherence: A Novel Nonpharmacological Modality for Lowering Blood Pressure in Hypertensive Patients. Global Advances in Health and Medicine, 2012. 1(2): p. 54-62.

4. McCraty, R., M. Atkinson, and D. Tomasino, Impact of a workplace stress reduction program on blood pressure and emotional health in hypertensive employees. J Altern Complement Med, 2003. 9(3): p. 355-69.

5. Alabdulgader, A., Coherence: A Novel Nonpharmacological Modality for Lowering Blood Pressure in Hypertensive Patients. Global Advances in Health and Medicine, 2012. 1(2): p. 54-62.

5b. Tai Chi research review. Field T. August 2011.

5c. Yoga and heart rate variability: A comprehensive review of the literature. Anupama Tyagi and Marc Cohen. Dec 2016,

5d. The neural mediators of kindness-based meditation: a theoretical model. Jennifer S. Mascaro. et al. Feb 2015.

5e. Additive effects of acetylcholine released by vagal nerve stimulation on arterial rate. Hondeghem LM et al. Jan 1975.

5f. Clitoria ternatea root extract enhances acetylcholine content in rat hippocampus. Rai KS et al. Dec 2002.

5g. Influence of clitoria ternatea extracts on memory and central cholinergic activity in rats. Taranalli AD and Cheeramkuzhy TC.

5h. Moderate pressure massage elicits a parasympathetic nervous system response. Diego MA and Field T. Int J Neurosci. 2009119(5):630-8. doi: 10.1080/00207450802329605.

Chapter 14. Acetylcholine its Effects upon Human Brainwaves

Theta Brainwaves and Acetylcholine

Theta brainwaves are also important to the success of ARV sessions. It is important to note here that the brain can and does have a various overlapping of frequencies. Studies conducted by Doronin et al., (1998) found that oscillations in the KP index and brainwave alpha-rhythms in people exhibited **identical** **periods**.

A study found that **theta** brainwave activity is associated with a burst-like discharge of **acetylcholine** within the basal forebrain [1]. Also the role of acetylcholine in the brain has been shown to enhance the intensity as well as the quality of neuron signaling by **increasing theta brainwave activity.** Acetylcholine also enhances the encoding of memories in the perirhinal and entorhinal cortex regions of the brain [2]. Speaking from personal experience, after having taken an extract of butterfly pea, I noticed that it induced extraordinarily vivid dreams, not to mention leaving me feeling extremely calm and relaxed for days afterwards. Hence acetylcholine may be a powerful way to enhance one's creativity.

Methods and Herbs that Enhance Acetylcholine Levels

The vagus nerve releases acetylcholine. Hence massaging the vagus nerves, which are situated just below the ear lobes, may create a release of acetylcholine.

Butterfly Pea, also called pigeonwings, bluebellvine, blue pea, Clitoria ternatea, cordofan pea and Darwin pea **enhances acetylcholine levels**. Butterfly Pea contains various triterpenoids, steroids, flavonol glycosides and anthocyanins

[3].

The herb Ashwagandha increases acetylcholine levels in the body [4].

Aged mice fed Convolvulus pluricaulis for 7 days were found to have better memory when given in the proper dosages [5].

White horehound has been found to potentiate acetylcholine release (El Bardai et al., 2001) [6].

The Full Moon

It may be that if stronger solar activity occurs during a full moon that a double shielding type effect takes place where the geomagnetic activity, no matter how strong, acts as a shield against more cosmic rays. Hence, less cosmic rays means less "fuel" for nervous system anticipatory responses leading to less accurate ARV sessions. This means even if geomagnetic activity is calm or low and there is a sudden increase in solar activity, the ARV session will be less accurate due to less cosmic rays. It is a fact that cosmic rays exhibit time dilation. On the other hand if a full moon is present, there are more cosmic rays and solar activity is minimal, ARV sessions tend to be more accurate.

I have also found over the last 10 years of practicing the Emerald Tablets exercise that when there are more cosmic rays and lower solar wind speeds, that the revitalization energies experienced during the exercise are much more stronger.

A long term study that took place when geomagnetic activity was quiet found strong positive relationships between cosmic rays and HRV variables [7]. This suggests HRV responds to increases in cosmic rays. Below is a quote from the study [7] -

'suggests that parasympathetic nervous system activity is enhanced during times of <u>increased</u> solar radio flux and <u>cosmic rays</u>.'

Summary

Enhanced cosmic rays are positively affecting HRV. This means there may exist a link between aging and cosmic rays. Because stronger solar activity acts as a shield against cosmic rays and stronger solar activity causes ill health effects, when solar activity is quiet or low, it allows more cosmic rays to enter earth, enhancing health and well being.

Although there are numerous studies showing cosmic rays are bad for health when a person is in outer space, I could find no studies showing that cosmic rays are bad for health for people living on the ground.

Cosmic rays follow a cycle. They rise when solar activity is quiet and they decline when solar activity is stronger. The following image is a chart showing the relationship between solar cycles and cosmic rays.

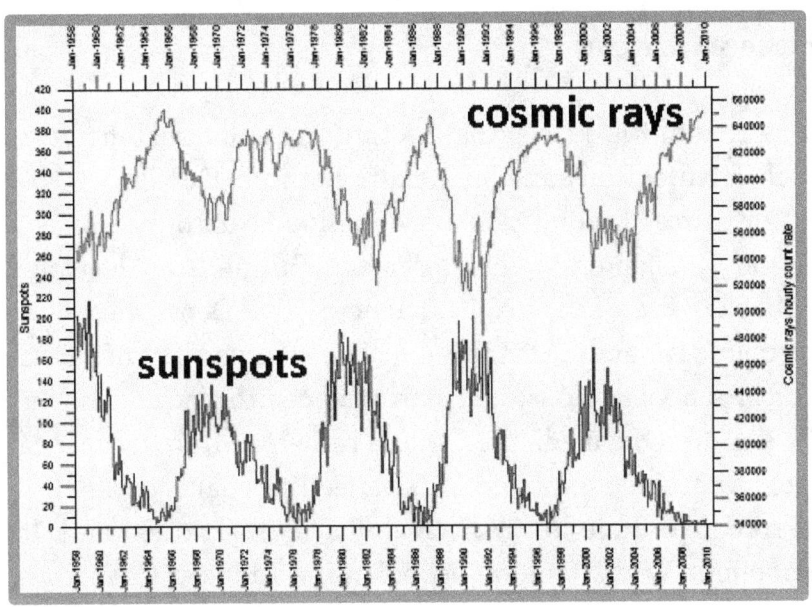

Above image courtesy of: Variation of Cosmic Ray Intensity and Monthly Sunspot Activity since 1958. The Germany Cosmic Ray Monitor in Kiel (GCRM) and NOAA's National Geophysical Data Center (NGDC). November 2009'

Summary

High sunspot activity correlates with low cosmic ray intensity, as well as vice versa.

The next chapter goes into more detail about the relationship between cosmic ray and HRV and, you can skip to Chapter 15 to learn more about the nervous system if you find you have had enough of learning about HRV.

References. Chapter 14.

1. The role of REM sleep theta activity in emotional memory. Isabel C. Hutchison1 and Shailendra Rathore. Oct 2015.

2. Waymire, J. Ph.D., Acetylcholine Neurotransmission (Section 1, Chapter 11) Neuroscience Online: An Electronic Textbook for the Neurosciences | Department of Neurobiology and Anatomy – The University of Texas Medical School at Houston. Neuroscience.uth.tmc.edu.

3. The Ayurvedic medicine Clitoria ternatea--from traditional use to scientific assessment. Mukherjee PK. et al. Dec 2008.

4. Protective Role of Ashwagandha Leaf Extract and Its Component Withanone on Scopolamine-Induced Changes in the Brain and Brain-Derived Cells.Arpita Konar et al. Nov 2011.

5. Ayurvedic medicinal plants for Alzheimer's disease: a review. Rammohan V. Rao et al. June 2012.

6. Anti-Hypertensive Herbs and Their Mechanisms of Action: Part IIM. Akhtar Anwar et al. March 2016.

7. Synchronization of Human Autonomic Nervous System Rhythms with Geomagnetic Activity in Human Subjects. Rollin McCraty et al. July 2017.

Chapter 15. HRV and related Parameters that Influence Coherence

A study looked at the body's serum C-reactive protein levels in people who were suspected of having inflammatory-related issues. The study discovered a strong / inverse correlation between their C-reactive protein levels and the intensity of cosmic rays [1]. In another published study the researchers observed that during the first 2-weeks, cosmic rays were weakly and negatively correlated to HRV measures. The study also found that the participants exhibited strong positive correlations with the LF/HF ratio measures and that a sharp increase in solar wind speeds was associated with a strong reduction in cosmic rays during the period in response to the existence of a moderate magnetic storm. Sharp increases in solar wind speeds are usually associated with a sudden increase in solar activity. The study concluded that cosmic rays and solar wind speeds affected all the key HRV measures [1b]. Hence it may mean that cosmic rays exhibit stronger effects on the body's autonomic nervous system compared to the sun's 10.7cm solar radio flux. If this is true, if cosmic ray activity has been increasing for 3 or more days, then it may be even more beneficial for ARV sessions than an increased 10.7cm radio flux. This would also mean that 3 or more days of increasing cosmic rays may override lower levels of the sun's 10.7cm radio flux during an ARV session. Also the solar wind speed during the study was negatively correlated with IBIs. This occurred when above average geomagnetic activity took place. IBIs also displayed an inverted relationship to the body's heart rate (HR) and showed longer IBIs in response to a slower HR (HR = 60/IBI).

What are HR and IBIs?

HR and IBIs are indicators for shifts that occur between the body's parasympathetic and sympathetic nervous systems. HR and IBIs also determine how the body's autonomic nervous system adapts or responds to challenges and stress [2].

This shows HR increases as solar wind speeds increase and also when above average geomagnetic activity occurs, causing physiological stress that exhibits carryover effects into the post-storm period. This effect is consistent with other studies [3] [4]. The study also found a positive correlation between the sun's solar radio flux (F10.7cm) and cosmic rays for the majority of the HRV variables. Also a negative correlation was found to occur between the LF/HF ratio during the beginning two weeks (unsettled period) of the study. This suggests that parasympathetic nervous system activity is enhanced during times of increased solar radio flux and when more cosmic rays are present.

It is a fact that of today's doctor visits, 60% to 80% are related to stress. What is alarming is that only 3% of patients receive any type of stress management help [5] [6] [7]. Hence, this shows that learning to go into coherence or practicing Associative Remote Viewing (ARV) can not only help increase your income, and reinforce belief in your intuition, but can lead to long term health benefits.

Physiological Coherence

This is determined by identifying maximum peaks occurring in the 0.04 to 0.26 hertz range of the HRV power spectrum and calculating the integral within a window of 0.030 hertzwide that is centered on the highest peak in that region and then to calculate the total power of the whole spectrum

(8).

A research study looking at the ECG of 11 patients that had Parkinson's disease were administered a dose of levodopa (to enhance dopamine levels) for deep brain stimulation treatment. HRV was recorded at three time points before the administration of the levodopa and 30 and 60 minutes after taking the levodopa. The study found that 30 minutes after taking the levodopa, that the participant's HRV parameters showed that their parasympathetic nervous system activity became decreased and that the sympatho-vagal balance was shifted towards sympathetic control. However, the interesting part of this study is that **60 minutes later** their parasympathetic nervous system became stimulated and a decrease in their heart rate occurred, suggesting a move towards a more active parasympathetic nervous system [9].

Final Conclusions

This study is consistent with other studies in that changes in geomagnetic and solar activity positively correlate with changes in the human nervous system. The study shows the human autonomic nervous system responds to changes in solar and geomagnetic activity and that it is efficiently synchronized with earth's time-varying magnetic fields. These fields are associated with Schumann resonances and geomagnetic field-line resonances. The sun's 10.7cm radio flux may be a major mediator of the human body's anticipatory reaction system, first observed by Chizhevsky.

What Does HRV Stand for?

Biological systems in sound health exhibit patterns that can be interpreted using mathematical abstracts. As a heart beats, small changes occur in-between each heart beat. Heart rate variability (HRV) measures these changes in real time and calls them IBI's (interbeat intervals). This is a similar system to the meditation practice where one learns to focus on the spaces between one's thoughts. Healthy hearts do not follow a distinct pattern. The beating of a healthy heart is complex, constantly changes and adapts to its surrounding environment. This allows on demand adjustments to be made by the cardiovascular system in order for it to adjust to sudden physical challenges and/or changes to homeostasis [9b].

Summary
HRV is a valuable and reliable indicator of a person's behavioural flexibility and psychological resiliency. HRV can also be used to measure how well a person has mastered self-mastery skills. This allows them to quickly adapt to changing environmental or societal demands [10] [11]. It allows one to gain control of their emotions when they feel overwhelmed, which in turn leads to positive changes in their body's physiological systems.

The Three Main Frequencies of HRV
HRV consists of 3 primary waveforms

HF
The high-frequency spectrum is the range between 0.15 and 0.4 hertz (rhythms with periods that take place between 2.5 and 7 seconds). This frequency represents a more active parasympathetic nervous system or more active vagal activity. It is commonly known as 'the respiratory band'. This is because it exhibits variations to HR fluctuations in response

to breathing.

Enhanced HF
In a healthy person high-frequency (HF) is increased by deep and slow breathing.

Lower HF
When the human cardiovascular system is oscillating at this frequency, it shows a high amplitude peak in the HRV power spectrum at around the frequency of 0.1 hertz. The 0.1 hertz frequency has a period of 10 seconds [12].

Physics equations demonstrate that the resonance frequency of the human cardiovascular system is governed by a series of feedback loops that exist between the brain and heart [13] [14]. In many mammals this frequency is approximately **0.1 hertz**, and exists as an overall coherent frequency.

Low HF is associated with panic, stress, anxiety and worry. It corresponds to reduced parasympathetic nervous system activity, rather than lower sympathetic functioning. It has been theorized that the LF band reflects sympathetic activity. It also accounts for higher ratios of reduced HRV in aging. Lower HF is also related to lower cognitive functioning.

The Low-Frequency Band
This band ranges from between 0.04 and 0.15 hertz (which is equal to rhythms with periods occurring between 7 and 25 seconds). This region was originally known as the baroreceptor range or the mid-frequency band range by researchers. This is due to the fact that it reflects baroreceptor activity while a person is at rest [15].

The Very-Low-Frequency Band
This is power in the HRV spectrum between 0.0033 and 0.04 hertz (which is equal to rhythms with periods that take place between 25 and 300 seconds). Increases in resting VLF power or a shifting of the frequency reflects efferent sympathetic activity. Very low VLF power is associated with an increased risk of arrhythmic death [16] and also PTSD (17). Also this band is associated with high inflammation [18] [19] and studies and have shown it is correlated with lower levels of testosterone. In a healthy person an increase in VLF power takes place during night-time and peaks before the person awakens [20] [21]. This increase appears to coincide with the morning cortisol peak. A normal and healthy VLF power is representative of healthy functioning.

HRV and Studies
A research study that involved a group of high school students that practiced self-regulation techniques over the course of four-months, found that their resting HRV was significantly increased. Also their pattern of HRV was significantly more coherent. The study also discovered that these increases resulted in improved behaviours as well as higher test scores. The study concluded that the practice of the self-regulation skills (self mastery) caused enhanced coherent heart rhythms which resulted in more coherent and stable rhythms in the students' cardiovascular and afferent neuronal networks [20]. This is a significant finding because it shows that education can be a major building block to learning how to apply self-regulation techniques.

References. Chapter 15

1. Dynamics of serum C-reactive protein (CRP) level and cosmophysical activity. Stoupel E Et al. March 2007.

1b. Long-Term Study of Heart Rate Variability Responses to Changes in the Solar and Geomagnetic Environment. Abdullah Alabdulgader et al. Feb 2018.

2. Shapiro, A.P., Hypertension and Stress: A Unified Concept1996, Mahwah, NJ: Lawrence Erlbaum Associates.

3. Goldstein, D.S., Stress, allostatic load, catecholamines, and other neurotransmitters in neurodegenerative diseases. Endocr Regul, 2011. 45(2): p. 91-8.

4. Siegman, A.W., et al., Dimensions of anger and CHD in men and women: self-ratings versus spouse ratings. J Behav Med, 1998. 21(4): p. 315-36.

5. Nerurkar, A., et al., When physicians counsel about stress: results of a national study. JAMA Intern Med, 2013. 173(1): p. 76-7.

6. Avey, H., et al., Health care providers' training, perceptions, and practices regarding stress and health outcomes. J Natl Med Assoc, 2003. 95(9): p. 833, 836-45.

7. Cummings, N.A. and G.R. Vanden Bos, The twenty years Kaiser-Permanente experience with psychotherapy and medical utilization: implications for national health policy and national health insurance. Health Policy Q, 1981. 1(2): p. 159-75.

8. Wölk, C. and M. Velden, Detection variability within the cardiac cycle: Toward a revision of the 'baroreceptor hypothesis'. Journal of Psychophysiology, 1987. 1: p. 61-65.

9. Autonomic nervous system response to L-dopa in patients with advanced Parkinson's disease. Ruonala V, et al. 2015.

9b. An Overview of Heart Rate Variability Metrics and Norms. Fred Shaffer and J. P. Ginsberg. Sept 2017.

10. Berntson, G.G., et al., Cardiac autonomic balance versus cardiac regulatory capacity. Psychophysiology, 2008. 45(4): p. 643-52.

11. Beauchaine, T., Vagal tone, development, and Gray's motivational theory: toward an integrated model of autonomic nervous system functioning in psychopathology. Dev Psychopathol, 2001. 13(2): p. 183-214.

12. Heart Rate Variability: New Perspectives on Physiological Mechanisms, Assessment of Self-regulatory Capacity, and Health risk. Rollin McCraty, PhD and Fred Shaffer, PhD, BCB. Jan 2015.

13. deBoer, R.W., J.M. Karemaker, and J. Strackee, Hemodynamic fluctuations and baroreflex sensitivity in humans: a beat-to-beat model. Am J Physiol, 1987. 253(3 Pt 2): p. H680-9.

14. Baselli, G., et al., Model for the assessment of heart period variability interactions of respiration influences. Medical and Biological Engineering and Computing, 1994. 32(2): p. 143 to 152.

15. Malliani, A., Association of Heart Rate Variability components with physiological regulatory mechanisms, in Heart Rate Variability, M. Malik and A.J. Camm, Editors. 1995, Futura Publishing Company, Inc.: Armonk NY. p. 173-188.

16. Bigger, J.T., Jr., et al., Frequency domain measures of heart period variability and mortality after myocardial infarction. Circulation, 1992. 85(1): p. 164-71.

17. Shah, A.J., et al., Posttraumatic stress disorder and impaired autonomic modulation in male twins. Biol

Psychiatry, 2013. 73(11): p. 1103-10.

18. Lampert, R., et al., Decreased heart rate variability is associated with higher levels of inflammation in middle-aged men. Am Heart J, 2008. 156(4): p. 759 e1-7.

19. Carney, R.M., et al., Heart rate variability and markers of inflammation and coagulation in depressed patients with coronary heart disease. J Psychosom Res, 2007. 62(4): p. 463-7.

20. Huikuri, H.V., et al., Circadian rhythms of frequency domain measures of heart rate variability in healthy subjects and patients with coronary artery disease. Effects of arousal and upright posture. Circulation, 1994. 90(1): p. 121-6.

21. Singh, R.B., et al., Circadian heart rate and blood pressure variability considered for research and patient care. Int J Cardiol, 2003. 87(1): p. 9-28 discussion 29-30.

20. Bradley, R.T., et al., Emotion self-regulation, psychophysiological coherence, and test anxiety: results from an experiment using electrophysiological measures. Appl Psychophysiol Biofeedback, 2010. 35(4): p. 261-83.

Chapter 16. The Autonomic Nervous System.

Our emotions, no matter how subtle they may feel to us, influence activity in our autonomic nervous system. Health problems arise due to improper functioning of the autonomic nervous system. The autonomic nervous system is particularly sensitive to intense emotions. For example, anger causes increased sympathetic activity and relaxation techniques increase parasympathetic activity.

HRV and Limonene

It is wonderful that nature has given us the gift of essential oils. What a wonderful way to reduce stress and at the same time enjoy such beautiful scents! What does the data have to say about this marvelous gift? A research study looked at the effectiveness of aromatherapy on participant's heart rate variability, blood pressure and the aortic augmentation index of essential hypertensive patients. The participants' were given a blend of oils of lavender (Lavandula angustifolia), lemon (Citrus limonum) and ylang ylang (Cananga odorata) which were mixed in the ratio of 2:2:1. A control group was given artificial Limonene (35 cc) and Citral (15 cc) mixture. All participants were told to inhale the mixtures twice daily for 3 weeks. The study found noticeable differences in the participants' systolic blood pressure values with a notable difference in their sympathetic nerve system activity of heart rate variability (p=.047). The study concluded that this combination of essential oils is effective in lowering a person's systolic blood pressure and reducing the activity of their sympathetic nervous system [1].

A few Quick Facts about the Autonomic Nervous System

The autonomic nervous system is influenced by our cardiovascular, digestive, immune, hormonal and other bodily networks.

Negative feelings or emotions can cause disorder in the autonomic nervous system.

Emotions such as appreciation create increased order in the autonomic nervous system. This causes improved immune and hormonal-system functioning as well as enhanced cognitive abilities.

Juniper Berry and the Autonomic Nervous System

From personal experience of using juniper essential oil, I have found that it smells very similar to cedar essential oil, but with its aroma being much more concentrated than cedar. Hence it is commonly sought after as a moth repellent. As a matter of fact juniper and cedar share a very close relationship to one another. Studies have found that juniper trees **live to almost 900 years** and **outlast cedar** in rot resistance. Juniper also allows plants to increase the efficiency of photosynthesis [2].

Juniper's Effect upon the Human Nervous System

A study looked at juniper essential oil's effect on the human autonomic nervous system. The study found that diastolic and systolic blood pressure became decreased when the participant's inhaled juniper essential oil. Also high frequency (HF) power level, which represents parasympathetic nervous system activity, was enhanced when the participant's inhaled

juniper essential oil. Also high frequency/low frequency (HF/LF) ratio, which indicates sympathetic nervous system status, was decreased by inhaling juniper essential oil. The study concluded that juniper essential oil effects a modulatory effect upon the human autonomic nervous system (Effects of Juniper Essential Oil on the Activity of Autonomic Nervous System Jong-Seong Park. 2017) Another study found that juniper berry oil from Bulgaria contained 51.4% pinene [5]. Juniper essential oil is also made up of the hydrocarbons limonene, thujone, myrcene and sabinene [6]. Another study found Juniper Essential Oil consisted of the following - a-pinene (27.22– 62.00%), limonene (1.31–30.96%), citronellol (5.06–15.57%), myrcene (5.41–20.23%) and sabinene (0.27–16.47%) [7].

Copaiba as an Essential Oil Amplifier

Research at the Young Living Essential Oil Lab found when Copaiba oil was mixed with Oregano and Peppermint essential oil, that the anti-inflammatory effect was **four times greater** [8]. An essential oil with similar anti-inflammatory effects is Ocotea [9], which may also enhance the effects of other essential oils, although research is necessary to confirm this. Copaiba is used as an anti-inflammatory in Brazilian folk medicine [10]. Hence, specific anti-inflammatory substances may exhibit amplification effects. Another reserach study found that Copaiba oil caused neuro-protection due to its effects in reducing inflammation due to acute damage in the central nervous system [10].

Not everyone knows about how they can incorporate coherence into their lives to make life a little calmer and less stressful. Perhaps some will not want to know because they may enjoy chaos. However in general things seem to go much better when one can integrate self-regulation techniques into their lifestyle.

Our universe operates on coherent principles which explains its orderly and endurant makeup. Coherence is found in multiple systems in nature. Its quasi-instant connection exists among parts of many things, whether it be an atom, a galaxy or an organism. This type of coherence exists in fields as diverse as biology, cosmology, quantum physics, the human brain and consciousness. Recent studies are discovering that all living systems are clearly interconnected at deep fundamental levels and that they have a communications system that occurs via biological fields and non-local mechanisms [11] [12].

Experiencing coherence leads to immediate benefits that transforms stress the moment coherence begins. Additional benefits include long-term emotional well-being and improvements in emotion regulation that over the long term positively affect one's life in numerous areas. This is due to the fact that people intentionally self-generate various states of physiological coherence rhythms. When the body's rhythms are coherent, the body responds better to challenges due to reinforced neural architecture. This comes from consistent practice, which allows for a re-patterning process to take place. This then creates a feed-forward process which becomes established via a new baseline or reference system, which the body strives to maintain.

Summary
Regular practice of heart / brain coherence reinforces the re-patterning process, where maladaptive patterns that cause

stress become replaced by healthier emotional, cognitive, physiological and behavioural patterns which become "automatic" or as a familiar way of being.

Physiological Coherence

This is the optimal prime state the body should be in for successful ARV sessions. Physiological Coherence occurs when a person is able to activate and sustain genuine feelings of deep appreciation. If one were to look at a person's physiological activity on a computer screen, emotions such as love or appreciation would be displayed as smooth, highly ordered sine-wave-like heart rhythms (also called coherence).

Benefits of the Heart Lock-In Technique

Research has found that an intentional activation of positive emotions is possible and that it can play important roles in increasing one's cardiac coherence, which then leads to self-mastery (self-regulatory capacity) [13].

A research study looked at EEG readings from 30 participants. The study involved looking at their physiological and baseline coherence modes. The Heart Lock-In technique was used to enter physiological coherence by allowing one to self-generate feelings of appreciation. As they did so, they listened to music that fostered positive emotions (Childre, 1991). The study found a significant increase in the participant's heart rhythm coherence in their physiological coherence and a major increase in their ECG–alpha synchronization in their brain's left hemisphere that was centred around the temporal lobe [14].

Mental Focus

Mental focus causes less synchronized activity to take place in the ANS, compared to relaxation or coherence. This is shown by erratic heart rhythm patterns [15]. The majority of structured mediation and relaxation techniques are disassociation techniques. However psychological states that encourage mind / body coherence have been scientifically related to activated positive emotions [15].

The Parasympathetic Nervous System

When a person experiences heart / brain coherence, their nervous system will display increased parasympathetic activity [14]. Increased parasympathetic activity causes large peaks to take place in the HF band of the power spectrum. Increased parasympathetic activity tends to take place during periods of relaxation and rest and even with structured meditation techniques. This typically causes an overall reduction in ANS outflow. When the body is at rest the parasympathetic and sympathetic nerves are tonically active, with vagal effects being predominant. In simple summary, coherence is associated with increased parasympathetic activity and is part of the relaxation response. Also meditation and relaxation do not always cause significant increases in coherence. This is because fundamental differences exist between coherence and the physiological correlates of relaxation. Also associated psychological states during these times can be markedly different. The vagus nerves, situated just under the ear lobes, are primary nerves for the human parasympathetic system and innervate the intrinsic cardiac nervous system.

References. Chapter 16.

1. Effects of aromatherapy on changes in the autonomic nervous system, aortic pulse wave velocity and aortic augmentation index in patients with essential hypertension. Cha JH1, Lee SH, Yoo YS. J Korean Acad Nurs. 2010 Oct40(5):705-13. doi: 10.4040/jkan.2010.40.5.705.
2. Biology and ecology of redberry juniper. Darrell N. Ueckert. Texas A&M.
3. Quantum effects in biology: golden rule in enzymes, olfaction, photosynthesis and magnetodetection. Jennifer C. Brookes. May 2017.
4. Effects of Juniper Essential Oil on the Activity of Autonomic Nervous System. Jong-Seong Park. Sept 2017. Department of Physiology, Chonnam National University Medical School, Gwangju 61469, Korea. Biomedical Science Letters 2017, 23(3): 286~289.
5. Chemical Composition and Antioxidant Properties of Juniper Berry (Juniperus communis L.) Essential Oil. Action of the Essential Oil on the Antioxidant Protection of Saccharomyces cerevisiae Model Organism.Höferl M et al. Feb 2014.
6. Antimicrobial activity of juniper berry essential oil (Juniperus communis L., Cupressaceae). Stjepan Pepeljnjak. Oct 2005.
7. Essential Oil of Juniperus communis L. Grown in Northern Greece: Variation of Fruit Oil Yield and Composition. Paul K. Koukos & Kaiti I. Papadopoulou. Dec 1995.
8. Copaiba. www..youngliving.com.
9. Antioxidant and Anti-inflammation Activities of Ocotea, Copaiba and Blue Cypress Essential Oils in Vitro and in

VivoR. Amilia Destryana. July 2014.

10. Copaiba Oil-Resin Treatment Is Neuroprotective and Reduces Neutrophil Recruitment and Microglia Activation after Motor Cortex Excitotoxic Injury. Adriano Guimarães-Santos. et al. Feb 2012.

11. Lane, R.D., et al., Activity in medial prefrontal cortex correlates with vagal component of heart rate variability during emotion. Brain and Cognition, 2001. 47: p. 97-100.

12. Reece, Jane B. (September 27, 2010). Campbell Biology (9 ed.). Benjamin Cummings. p. 205. ISBN 978-0-321-55823-7.

13. McCraty, R., Atkinson, M., Tomasino, D., & Bradley, R. T, The coherent heart: Heart-brain interactions, psychophysiological coherence, and the emergence of system-wide order. Integral Review, 2009. 5(2): p. 10-115.

14. The Coherent Heart Heart–Brain Interactions, Psychophysiological Coherence, and the Emergence of System-Wide Order Rollin McCraty. et al. Dec 2009.

15. Siegman, A.W., et al., Dimensions of anger and CHD in men and women: self-ratings versus spouse ratings. J Behav Med, 1998. 21(4): p. 315-36.

Chapter 17. Coherent and Incoherent States of Being

Coherence does not always mean a change in a person's heart rate or a change in the amount of heart rate variability (HRV). Rather, it takes place as a distinctive shift in a person's heart rhythm pattern. This physiological state is usually associated with above average high-amplitude peaks in the LF band which are centered around approximately 0.1 Hz. This shows that system-wide resonance is taking place. Also it means stronger synchronization is occurring between the sympathetic and parasympathetic regions of the human nervous system, as well entrainment between heart and blood pressure rhythms.

Hence heart rhythm in relaxation and coherence appears as a sine-wave-like pattern. In physiological coherence mode, this pattern shows up as a lower frequency and usually with higher amplitude. Also coherence shows increased resonance, synchronization, and entrainment across multiple bodily systems. These represent global organizations not present in the standard experience of relaxation.

Benefits of Coherence

Before we can get into using coherence to enhance our Associative Remote Viewing Sessions, we first need to understand how a coherent heart impacts the body. Being a good typist, I am always seeking ways to type well and fast. Major hurdles to typing include sluggish speed and inaccuracy. Wouldn't it be great if coherence / self-regulation techniques could enhance one's accuracy in tasks involving clerical type activities? Feedback from individuals practicing coherence exercises stated that their performance in various activities had shown major improvements. These included abilities that required the processing of various types of external sensory information such as coordination, speed, accuracy and synchronization. These types of sensory activity take place in various sports and include the performing arts. These types of activities require techniques that generate strong internal focus to help with decision making, problem solving, creativity and intuition. Also these attributes reveal themselves in intellectual and business environments. This means that physiological bodily coherence and an organized pattern of the heart has positive effects on the brain's cognitive processes and a person's intentional behaviour.

Coherence Improves Cognition

An independent study conducted by Dr. Keith Wesnes in the UK found that practicing HeartMath positive emotion coherence enhanced the participant's memory capacity and that it also improved a person's feelings of calmness. Remote Viewing and Associative Remote Viewing are both processes that involve cognitive processing and in many cases more intensive cognitive functions are utilized during the session. Hence the better equipped the brain is to receive information, the better the mind is able to overcome any obstacles.

Another study that was conducted by the U.S. Department of Education at the Claremont Graduate University's School of Educational Studies included tenth grade students from two California high schools. The study discovered higher test scores and a significant reduction in test anxiety for students who practiced the positive emotion-focused coherence-building method (The TestEdge program) [1].

Solar Weather and Violence

The most interesting finding we have observed over the years is that when solar weather conditions are unfavourable, it results in more violence and ill health. Hence this may be due to solar weather causing disruption and interference in the body's physiological systems. This effect is described in greater detail in the book Solar Flares and Their Effects upon Human Behavior and Health. Geomagnetic cycles have not only been shown to correlate with human health indicators, but they also correlate with major societal conflicts such as terrorism, violence, crime and war. This finding has also been discovered by the Solar Institute where our research showed that when geomagnetic activity was at above average levels that terrorist attacks were much more common (www.ez3dbiz.com/earth_behavior.html). Research studies conducted by Halberg and Persinger found crime and war and were correlated to GMA [2].

Individuals experience feelings of enhanced well-being during coherence. This is due to the reduced inner "noise" generated by the emotional and mental processing of daily stress and due to the positive emotion-driven shift towards enhanced harmony in bodily processes. Many also report increased efficacy and intuitive clarity. A study of Zen monks discovered that the more advanced monks had coherent heart rhythms, while the novices did not (Lehrer et al., 2003).

Many people are unaware that the very ground they walk upon emits fields of energy which can lead to health problems. This takes place in the form of a geomagnetic storm. A great deal of well researched studies confirm that geomagnetic storms are related to death and human health [3] [4]. Altered EEG rhythms have been found to take place in studies conducted by Belov et al [5]. Also low-frequency magnetic oscillations (at approximately 3 hertz) have sedative-like effects and stronger oscillations of approximately 10 hertz have been found to stress and stimulate or motivate people (Pobachenko et al). [5]. This is a major finding because peaking geomagnetic activity has a motivational type effect and could explain why terrorist attacks occur more often when geomagnetic activity peaks.

One of the main contributors to ill health from higher solar activity is a significant increase in the incidence of heart attacks and myocardial infarction incidence [6].

Depression and Geomagnetic Storms

Other studies have found a 30% to 80% increase took place in hospital admissions for suicides, homicides, cardiovascular death, depression, mental disorders, cardiovascular disease, psychiatric admission and traffic accidents [7 to 12]. An independent research study undertaken by Pobachenko et al. found that birth-rates dropped and mortality rates rose during increased geomagnetic and solar activity (GMA). Also migraine attacks were triggered [13].

The very first report of an association between geomagnetic storms and psychiatric disturbances was reported by Dull & Dull (1935). Their study shows that increased psychiatric disturbances would occur between two or three days **after the geomagnetic storm** had passed. Hence, a 3 day delay exists in which physical and / or mental disturbances manifest themselves after the initial onset of a geomagnetic storm. Many of these negative health conditions take place more often during the spring and fall equinoxes, which are months geomagnetic energy peaks. The following image shows the sseasonal variations of intense geomagnetic storms for the years 1965–1975, 1976–1986, 1987–1997 and 1998–2008.

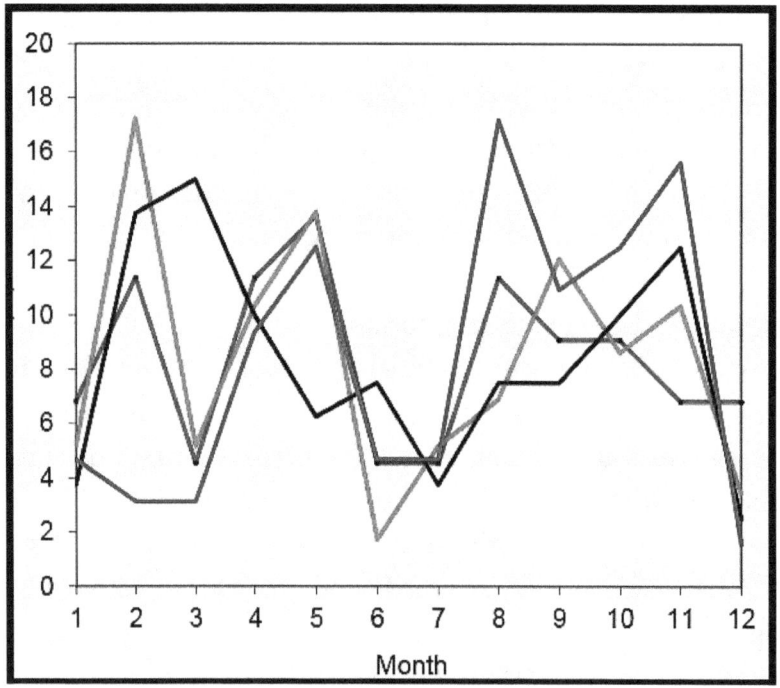

Above image courtesy of Solar Cycle and Seasonal Distribution of Geomagnetic Storms with Sudden Commencement Gustavo A. Mansilla et al. Dec 2013.

As the next image shows, treatment for depression happens to be more common during the 2 equinoxes.

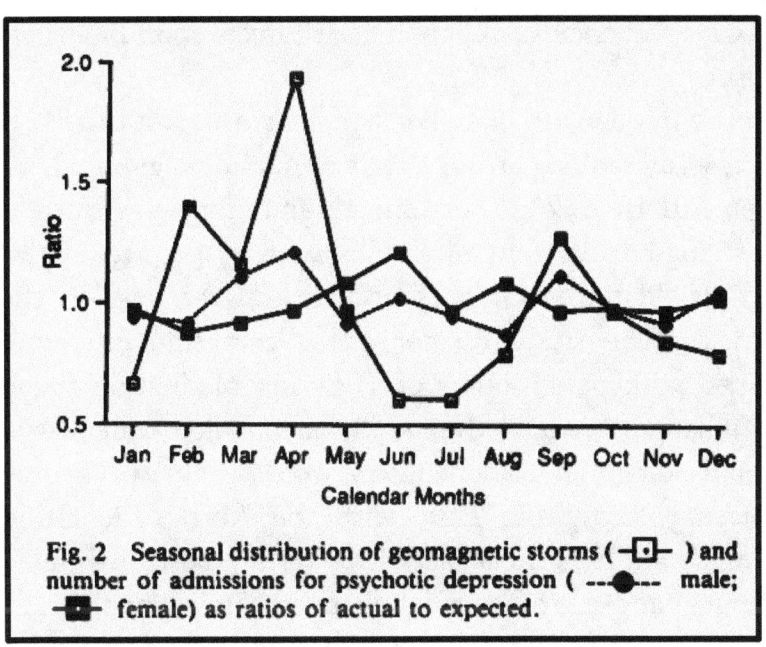

Fig. 2   Seasonal distribution of geomagnetic storms (–◻–) and number of admissions for psychotic depression (--●-- male; –■– female) as ratios of actual to expected.

Above image courtesy of Geomagnetic Storms: Association With Incidence of Depression as Measured by Hospital Admission. Kay RW. March 1004.

The above study found that geomagnetic disturbances associated with psychotic depressive illness occurs much more often in males and less in females.

Solar Wind Speed and Hallucinations

Randall and Randall (1991) studied data from the 19th century on hallucinations and magnetic disturbances and discovered a direct and statistically significant correlation between the solar wind speed and hallucinations. Hence, we see a confirmation that geomagnetic activity and remote viewing are sensitive to changes in the solar wind speed.

Ultra Low Frequencies (ULF) and their Effects upon Biological Organisms

One of the frequencies of a geomagnetic storm is 0.1hz [14]. A research study looking at the effects of pulsed magnetic fields between 0.01 Hz to 20 Hz on animals and humans with their heads facing in different directions with respect to earth's magnetic field found changes occurred in response to the frequencies. The study further stated that similar studies conducted with practitioners of Yoga and Meditation found that changes in the person's neurochemistry, electrophysiology and biochemistry would occur when they were facing North and East, with the North orientation reducing brain electrical activity. They also exhibited biochemical and neurochemical changes. However the East facing subjects were calmer, more alert and experienced more bliss [15]. Maybe there is something to Feng Shui after all?

Further Reading

The Effect of Extremely Low Frequency Alternating Magnetic Fields on the Behavior of Animals in the Presence of the Geomagnetic Field. Natalia A. Belova and Daniel Acosta-Avalos. Dec 2015.

Speaking from personal observation, the Emerald Tablets exercise involves laying down with the head facing north and when solar weather is favorable, the revitalization energies are significantly stronger. It may be that a healthy tension exists when facing north that enhances the strength of the mind. Let's next look at solar weather and its specific effects upon the nervous system and other systems of the body in the next chapter.

References. Chapter 17.

1. Bradley, R.T., et al., Reducing Test Anxiety and Improving Test Performance in America's Schools: Results from the TestEdge National Demonstration Study2007, Boulder Creek, CA: HeartMath Research Center, HeartMath Institute, Publication No. 07-09-01.

2. Persinger, M.A., Wars and increased solar-geomagnetic activity: aggression or change in Intra-species dominance? Percept Mot Skills, 1999. 88(3 Pt 2): p. 1351-1355.

3. Persinger, M.A., Sudden unexpected death in epileptics following sudden, intense, increases in geomagnetic activity: prevalence of effect and potential mechanisms. Int J Biometeorol, 1995. 38(4): p. 180-187.

4. Stoupel, E., Sudden cardiac deaths and ventricular extrasystoles on days of four levels of geomagnetic activity. J. Basic Physiol. Pharmacol., 1993. 4(4): p. 357-366.

5. Belov, D.R., Kanunikov, I. E., and Kiselev, B. V., Dependence of human EEG synchronization on the geomagnetic activity on the day of experiment. Ross Fiziol. Zh Im I M Sechenova, 1998. 84(8): p. 761–774.

6. Villoresi, G., Ptitsyna, N.G., Tiasto, M.I. and Iucci, N., Myocardial infarct and geomagnetic disturbances: analysis of data on morbidity and mortality [In Russian]. Biofizika, 1998. 43(4): p. 623-632.

7. Kay, R.W., Geomagnetic Storms: Association with Incidence of Depression as Measured by Hospital Admission. British Journal of Psychiatry, 1994. 164: p. 403-409.

8. Gordon, C., Berk, M. , The effect of geomagnetic storms on suicide. South African Psychiat. Rev, 2003. 6: p. 24-27.

9. Kay, R.W., Schizophrenia and season of birth: relationship to geomagnetic storms. Schiz Res, 2004. 66: p. 7-20.

10. Malin, S.R.C.a.S., B.J., Correlation between heart attacks and magnetic activity. Nature, 1979. 277: p. 646-648.

11. Nikolaev, Y.S., Rudakov, Y.Y., Mansurov, S.M. and Mansurova, L.G., Interplanetary magnetic field sector structure and disturbances of the central nervous system activity. Reprint N 17a, Acad. Sci USSR, IZMIRAN, Moscow, 1976: p. 29.

12. Oraevskii, V.N., Breus, T.K., Baevskii, R.M., Rapoport, S.I., Petrov, V.M., Barsukova, Zh.V., Gurfinkel' IuI, and Rogoza, A.T. , Effect of geomagnetic activity on the functional status of the body. Biofizika, 1998. 43(5): p. 819-826.

13. Zaitseva, S.A.a.P., M. I., Effect of solar and geomagnetic activity on population dynamics among residents of Russia [In Russian]. Biofizika, 1995. 40(4): p. 861-864.

14. Low Frequency F-Geomagnetic Fluctuations (0.025 To 20 Hz) on The Floor of Monterey Bay. Morgan P. Ames, Jr. and Dtic Louis Mcbane Vehslage. December 1981.

15. Effect of magnetic micropulsations on the biological systems — A Bioenvironmental Study. S. Subrahmanyam, et al. Jan 1985.

Chapter 18. Mechanisms of Solar Weather and its Impacts upon The Body.

Because solar activity, especially higher geomagnetic activity has a motivating and stimulating effect, there exist cycles in human history where great accomplishments in sports, arts, science, architecture, and positive social change have taken place [1]. This shows that solar activity exerts a sort of driving force upon the collective unconscious.

Summary
Out of all bodily systems studied and researched and their effects from changes in solar and geomagnetic activity, it is the rhythms of the brain and heart that appear to be most strongly affected [2 to 11]. The earth and its ionosphere generate a harmonious symphony of varying resonant frequencies which directly overlap with frequencies of the human brain and its cardiovascular system. Changes in these resonances, which can occur from varying solar weather, influence the function of the human autonomic nervous system, cardiovascular system and brainwaves.

Solar Weather's Effect upon the Human Nervous System
Early on in our remote viewing sessions we tended to look months or even weeks out into the future. This turned out to be a disaster. The way we solved this was due to the fact that solar activity follows a predictable cycle that averages 4 days of favourable solar weather conditions. If solar weather turns unfavourable up until the final date being remote viewed, ARV sessions become completely inaccurate. When solar weather conditions are favorable, we have successfully drawn

pictures of the trading activity of the Dow Jones and FOREX currencies anywhere from 1 to 4 days in advance. Because creativity / artistry is closely connected to the nervous system, the nervous system must be playing a vital / key role in the clarity of ARV sessions. It may be that the nerves of the nervous system are behaving as tiny antennas whose signals they receive have a limited range of just 4 days for detailed information and perhaps an unlimited range for major events such as earthquakes, major market crashes and other events that have a large scale impact upon human civilization.

As we have shown throughout this book, the main explanation for describing how geomagnetic and solar activity influence the human nervous system is by a resonant coupling between the human nervous system and geomagnetic frequencies (Alfvén waves), which are also known as Schumann resonances. Schumann resonances are better explained as ultra low frequency standing waves. These low frequency waves take place in the earth-ionosphere resonant cavity by multiple lightening strikes which overlap with the body's physiological rhythms [12].

Geomagnetic Storms, Solar Activity and their Impact on HRV

Increased heart rate activity (HR), a reduction in heart rate variability (HRV) and an increased number of arrhythmic events have been shown to take place during magnetic storms [13]. It may be that disruptions in environmental magnetic fields stress the body's physiological systems which in turn trigger changes in the brain's electrical activity [14]. The mechanisms that most likely explain how geomagnetic and solar influences affect health and behaviour are couplings that exist between the human nervous system and resonating geomagnetic frequencies known as Schumann resonances which take place in Earth's ionosphere resonant cavity [15].

Geomagnetic Activity and the Nervous System

A study looked at relationships between geomagnetic and solar activity and how it affected the human nervous system. The study involved 10 participants over 1 month whose physiological systems were continuously recorded in the form of HRV data [16]. The measuring equipment looked at time-varying changes in local geomagnetic, solar and Schumann resonance activity. The month long study also found steep increases in the participant's HRV after increased solar radio flux activity. Their **HRV declined rapidly** in conjunction with a **sharp jump in the solar wind** speed on 26th September, 2016 which caused a severe magnetic storm that started at approximately 12:15 universal time on 26 September, 2016 (NOAA Kp index of 8).

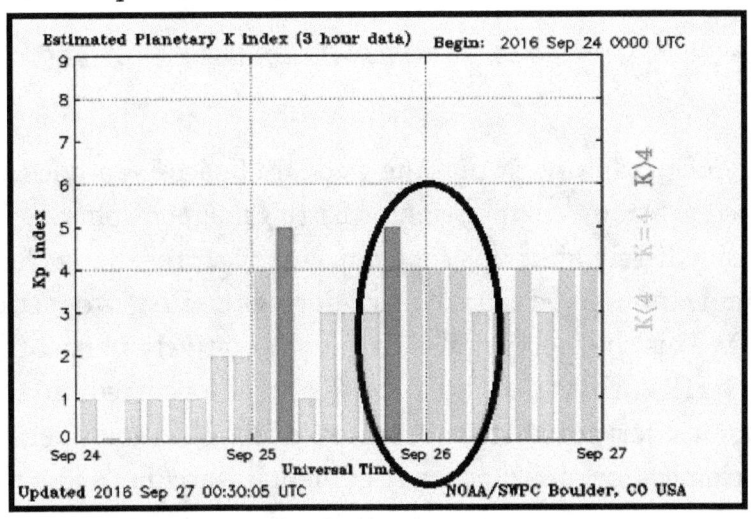

Referemce for previous image

http://legacy-www.swpc.noaa.gov/ftpdir/warehouse/2016/2016_plots/kp/20160926_kp.gif

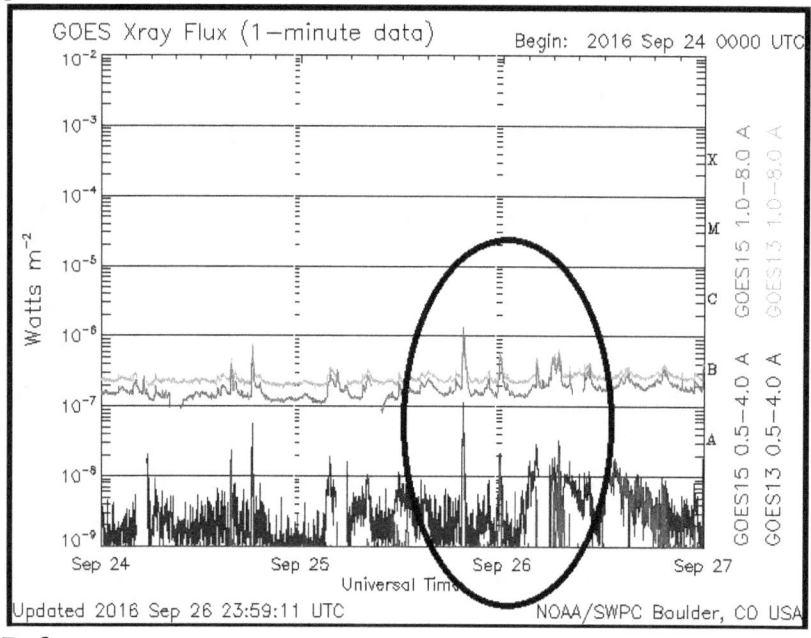

Reference:

http://legacy-www.swpc.noaa.gov/ftpdir/warehouse/2016/2016_plots/xray/20160926_xray.gif

It was during this event that the Goddard Space Weather Lab reported a "strong compression of Earth's magnetosphere". It was immediately after this severe magnetic storm that a clear rhythm was observed to take place during the first two weeks of SRP. This rhythm was also found to negatively impact the group's HRV rhythm. The time lags that occurred in the autonomic nervous system occurred over a 40-hour period after changes occurred in environmental variables, and for the changes in SRP, positive responses lagged increases in

Schuman resonance power by approximately 9 hours. The study concluded a **positive relationship exists between the Schuman resonance power and HRV.** In addition, this rhythm turned negative during times of magnetic disturbances. When the magnetic field disturbances ended, it was found that the group's HRV rhythms re-synchronized to a similar fashion first observed during the first two weeks of the study period [16].

Summary

The Schuman resonance power (SRP) and HRV are well-correlated and display clear repeatable oscillatory patterns. This shows environmental magnetic fields cause changes in the body's autonomic nervous system. **During periods of quiet geomagnetic activity, the effects of the Schuman resonance power cause a synchronization to take place in people's slow wave heart rhythms.**

Additionally, the human body's daily autonomic nervous system activity responds to changes in geomagnetic and solar activity and also synchronizes with time-varying magnetic fields that are associated with geomagnetic field-line (vector) resonances and earth's Schumann resonances. The properties that have the most impact upon the body are the Schuman Resonance and earth's magnetic field, and the sun's **solar wind speed.**

Schuman Resonance Synchronization

During the late 1950s, Schumann and Koenig measured specific frequencies that were consistent with a mathematical model predicting an earth-ionospheric cavity resonance [19]. They found the frequency of the first Schumann resonance

(SR) to be approximately 7.83 Hz with a day/night variation of approximately ±0.5 Hz. The other SR frequencies are ~14, 20, 26, 33, 39, and 45 Hz. These frequencies closely overlap with active human brainwaves. These human brainwave frequencies include alpha (8–12 Hz), beta (12–30 Hz) and gamma (30–100 Hz). Koenig [20] discovered that the close similarity between the frequencies produced by the human brain and earth's Schuman resonance caused the body's electroencephalogram rhythms to become synchronous with one other.

Pobachenko et al [21] monitored the SR's and EEG's of 15 individuals over six weeks and found that variations in their EEG's correlated with daily changes in the Schuman resonance and that the largest changes between their EEG's and earth's Schuman resonance occurred during periods of above average magnetic activity (geomagnetic storms). Studies since conducted by Persinger et al. looking at the Schuman resonance and EEG activity in real-time, showed several of the Schuman resonance frequencies are evident in the spectral profiles of human brain activity [22] [23]. Further studies yielded that the power within the brain's EEG spectral profiles demonstrated repeated periods of coherence with the first three SR resonance frequencies. These frequencies were 7–8 Hz, 13–14 Hz, and 19–20 Hz respectively. This coherence was found to take place in real-time. (Similar Spectral Power Densities Within the Schumann Resonance and a Large Population of Quantitative Electroencephalographic Profiles: Supportive Evidence for Koenig and PobachenkoKevin S. Saroka. Jan 2016).

Further Reading

Human Quantitative Electroencephalographic and Schumann Resonance Exhibit Real-Time Coherence of Spectral Power Densities: Implications for Interactive Information Processing. Michael A. Persinger and Kevin S. Saroka. May 2015.

Summary

A coherent standing wave takes place between human brainwaves and the first three frequencies of earth's Schuman resonance. Changes in the earth's Schuman resonance, caused by changes in solar activity, especially that of the solar wind and solar radiation are impacting brainwave activity, causing modulations in human memory consolidation and cognition [23].

This acts as a double edged sword. When geomagnetic activity is above average it contributes to ill health. When solar wind speed is at favourable levels (350 solar wind speed) and geomagnetic activity is low, it contributes to good health. Hence practicing HeartMath healing exercises during the good periods of solar activity can lead to rapid healing and maybe even reverse aging. This is because an enhanced Schuman resonance has been shown to affect the body's blood pressure by lowering it [24]. High blood pressure is one of the leading contributors to ill health.

Benefits of an Enhanced Resonance

We all get up in the morning and feel great after a shower. By the afternoon we don't feel as refreshed and maybe even have a little less energy. A mini-solar cycle that goes from quiet to active to quiet again lasts approximately 11 days on average works the same way. The energy builds, reaches a peak and then fades. It is the fading process that is key to successful ARV sessions. It may be that after a period of major / increased solar activity and geomagnetic activity, a 'fading' of the high energies begins. As this activity calms down a 'renewed' or 'refreshed' Schuman resonance takes place, kind of like hitting the reset key on your cell phone. Hence this cycle of activity may be acting as a cleansing mechanism for earth's ionosphere / magnetosphere. The study shown earlier showed SRP (Schuman resonance power) was strongly and positively associated with increased HRV across all IBIs (lower heart rate) and measures. Hence, it may also be conductive to various forms of health, especially if a person's diet consists of the right healthy foods.

Summary
A beneficial effect occurs during increased SRP (Schuman resonance power). This effect has been found to occur in other studies that display reduced systolic, diastolic, and mean blood pressure during periods when the SRP was higher [25].

The Schuman Resonance and its Interaction with the Coherent Heart
A research study involving experienced Buddhist monks that meditated on loving-kindness and compassion discovered that their heart rhythms showed increased coherence during their practice (Rapgay, n.d.) [26]. Conscious input signals originating from the brain to the heart affects activity of the brain's thalamus including its ability to synchronize cortical activity. This allows for stronger emotional stability, global

coherence and optimal performance [27].

Final Conclusions and Summary. Space Weather and its Effects on HRV

Significant autonomic nervous system responses affect HRV in response to cosmic ray activity. Schumann Resonance Power is highly correlated with cosmic ray counts. The solar radio flux causes robust responses in TP, IBIs, VLF and HF power. Increased Polar Cap activity causes significant negative correlations with TP at 7 hours and IBI at 8 hours. Daily ANS activity, reflected by HRV measures is reacting to changes in solar and geomagnetic activity during periods of normal undisturbed activity. ANS responses take place at different times after changes in the environment factors which persist over different lengths of time. Different individuals respond differently to these changes in the same environments. Solar wind speed is highly correlated with Ap, Kp and the PC(N) which reflect magnetic field disturbances. ULF power, which relates to magnetic field-line resonances, is positively correlated with the sun's solar wind speed. Indices of field disturbance are negatively correlated with cosmic ray counts. This is consistent with the well-known inverse action of geomagnetic and solar activity and cosmic ray counts at the Earth's surface. Positive correlations exist between HF power and solar radio flux. This indicates enhancements of the parasympathetic nervous system during periods of increased solar radio flux.

A study consisting of 1,643 participants in 51 countries discovered the 10.7cm solar radio flux index was positively correlated with improved emotional positive effects, reduced fatigue and enhanced mental clarity [28]. Also the same study also found that increases in solar wind speed had the opposite effects. Other effects caused by the solar radio flux showed an effect on the rate of death and found a inverse and strong relationship between the F10.7cm and rates of death (Stoupel E, et al, 2011, 2006). Hence the sun's 10.7 cm solar radio flux may be a mediator of pre-sentiment activity, especially the time lag which exists several days before increases in the solar wind from solar flares before reaching Earth, causing magnetic disturbances (geomagnetic storms).

Responses of the ANS to increases in cosmic rays showed that the largest ANS relationships were to cosmic rays. This supports the theory of Stoupel that cosmic rays are a principal factor of the environmental forces affecting human physiology [29].

Schumann Resonance power

Another environmental variable strongly associated with increased HF, LF and VLF power and TP of HRV measures is Schumann Resonance Power (SRP). This is accompanied by positive correlations to IBIs (lower heart rate). Beneficial effects of enhanced SRP is reduced diastolic, systolic and mean arterial blood pressure [30].

In summary, energetic environmental factors either indirectly or directly affect human psychophysiology and behaviors in different numerous ways which are dependant upon the health status, resilience and maturity of individuals [31].

This theory is supported by the fact that cosmic rays, increases in the sun's 10.7cm solar radio flux and Schumann Resonance Power are associated with increased parasympathetic activity and HRV. The ANS responds quickly to changes in Schumann resonance power, cosmic rays and the solar radio flux. In society in general, these increases are the prime motivators for increased social unrest, motivation and human flourishing. Also ANS responses change at different times after changes in various environmental factors and continue on for varying lengths of time.

Because solar wind is negatively correlated with IBIs, which indicates that the heart rate increases with the sun's solar wind speed, it suggests a physiological stress reaction occurs [32] [33].

Another observation made by the Solar Institute over the years is that when there is moderate to low solar activity with a small number of sunspots, it acts as an amplifier. Hence if the solar wind speed is around 450 or above, it will cause interference during the ARV session. If the solar wind speed is around 350 and there is moderate sunspot activity it will greatly enhance the clarity of the ARV session, especially if the moon is full.

References. Chapter 18.

1. Ertel, S., Cosmophysical correlations of creative activity in cultural history. Biophysics, 1998. 43(4): p. 696-702.

2. Halberg, F., et al., Cross-spectrally coherent ~10.5- and 21-year biological and physical cycles, magnetic storms and myocardial infarctions. Neuroendocrinology, 2000. 21: p. 233-258.

3. Pobachenko, S.V., Kolesnik, A. G., Borodin, A. S., Kalyuzhin, V. V., The Contigency of Parameters of Human Encephalograms and Schumann Resonance Electromagnetic Fields Revealed in Monitoring Studies. Complex Systems Biophysics, 2006. 51(3): p. 480-483.

4. Persinger, M.A., Sudden unexpected death in epileptics following sudden, intense, increases in geomagnetic activity: prevalence of effect and potential mechanisms. Int J Biometeorol, 1995. 38(4): p. 180-187.

5. Oraevskii, V.N., Breus, T.K., Baevskii, R.M., Rapoport, S.I., Petrov, V.M., Barsukova, Zh.V., Gurfinkel' IuI, and Rogoza, A.T. , Effect of geomagnetic activity on the functional status of the body. Biofizika, 1998. 43(5): p. 819-826.

6. Halberg, F., et al., Cycles Tipping the Scale between Death and Survival (="Life"). Progress of Theoretical Physics Supplement 2008. 173: p. 153-181.

7. Otsuka, K., et al., Chronomics and "Glocal" (Combined Global and Local) Assessment of Human Life. Progress of Theoretical Physics Supplement, 2008. 173: p. 134-152.

8. Persinger, M.A., Geopsychology and geopsychopathology: Mental processes and disorders associated with geochemical and geophysical factors. Experientia, 1987. 43: p. 92-104.

9. Dimitrova, S., Stoilova, I. and Cholakov, I., Influence of

Local Geomagnetic Storms on Arterial Blood Pressure. Bioelectromagnetics, 2004. 25: p. 408-414.

10. Hamer, J.R., Biological entrainment of the human brain by low frequency radiation. Northrop Space Labs, 1965: p. 65-199.

11. Rapoport, S.I., Malinovskaia, N.K., Oraevskii, V.N., Komarov, F.I., Nosovskii, A.M. and Vetterberg, L., , Effects of disturbances of natural magnetic field of the Earth on melatonin production in patients with coronary heart disease. Klin Med (Mosk), 1997. 75(6): p. 24-26.

12. Synchronization of Human Autonomic Nervous System Rhythms with Geomagnetic Activity in Human Subjects. Rollin McCraty et al. July 2017.

13. Armour, J.A., Potential clinical relevance of the 'little brain' on the mammalian heart. Exp Physiol, 2008. 93(2): p. 165-76.

14. Armour, J.A., Neurocardiology--Anatomical and functional principles. 2003, Boulder Creek, CA: HeartMath Research Center, HeartMath Institute, Publication No. 03-011.

15. Human Heart Rhythm Sensitivity To Earth Local Magnetic Field Fluctuations. Abdullah Alabdulgade, Rollin Maccraty, Mike Atkinson, Alfonsas Vainoras, Kristina Berškiene, Vilma Mauriciene, Alge Daunoraviciene, Et Al. 3272 © Jve International Ltd. Journal Of Vibroengineering. Sep 2015, Volume 17, Issue 6. Issn 1392-8716 .

16. Synchronization of Human Autonomic Nervous System Rhythms with Geomagnetic Activity in Human Subjects. Rollin McCraty et al. July 2017.

17. Crooker, N. Feynman, J. Gosling, J. On the high correlation between long-term averages of solar wind speed

and geomagnetic activity. J. Geophys. Res. 1977, 82, 1933–1937.

18. McPherron, R.L. Magnetic pulsations: Their sources and relation to solar wind and geomagnetic activity. Surv. Geophys. 2005, 26, 545–592.

19. Schumann, W. Konig, H. Uber die beobachtung von "atmospherics" bei geringsten frequenzen. Die Naturwiss. 1954, 41, 183–184.

20. König, H.L. Krueger, A.P. Lang, S. Sönning, W. Biologic Effects of Environmental Electromagnetism Springer: Berlin, Germany, 2012.

21. Pobachenko, S.V. Kolesnik, A.G. Borodin, A.S. Kalyuzhin, V.V. The contigency of parameters of human encephalograms and Schumann resonance electromagnetic fields revealed in monitoring studies. Complex Syst. Biophys. 2006, 51, 480–483.

22. Saroka, K.S. Persinger, M.A. Quantitative evidence for direct effects between earth-ionosphere Schumann resonances and human cerebral cortical activity. Int. Lett. Chem. Phys. Astron. 2014, 20, 166.

23. Persinger, M.A. Saroka, K.S. Human quantitative electroencephalographic and Schumann resonance exhibit real-time coherence of spectral power densities: Implications for interactive information processing. J. Signal Inf. Process. 2015, 6, 153.

24. Does Schumann resonance affect our blood pressure?G. Mitsutake et al. March 2009.

25. Mitsutake, G. Otsuka, K. Hayakawa, M. Sekiguchi, M. Cornélissen, G. Halberg, F. Does Schumann resonance affect our blood pressure? Biomed. Pharmacother. 2005, 59, S10–S14.

26. The embodied mind: A review on functional genomic and neurological correlates of mind-body therapies. Muehsam D. Feb 2017.

27. Synchronized activities of coupled oscillators in the cerebral cortex and thalamus at different levels of vigilance. Steriade M. Sept 1997.

28. Long-Term Study of Heart Rate Variability Responses to Changes in the Solar and Geomagnetic Environment. Abdullah Alabdulgader et al. Feb 2018.

29. Stoupel E, et al. Twenty years study of solar, geomagnetic, cosmic ray activity links with monthly deaths number (n-850304) Journal of Biomedical Science and Engineering. 20114:426. doi: 10.4236/jbise.2011.46054.

30. Does Schumann resonance affect our blood pressure? Mitsutake G et al. Biomed Pharmacother. 2005 Oct 59 Suppl 1():S10-4.

31. McCraty, R. &Deyhle, A.In Bioelectromagnetic and Subtle Energy Medicine, Second Edition (ed. Paul, J. Rosch) 411–425 (CRC Press 2015).

32. Long-Term Study of Heart Rate Variability Responses to Changes in the Solar and Geomagnetic Environment. Abdullah Alabdulgader et al. Feb 2018.

33. The global coherence initiative: creating a coherent planetary standing wave. McCraty R, Deyhle A, Childre D. Glob Adv Health Med. 2012 Mar 1(1):64-77.

Chapter 19. Additional Techniques that Create Heart /Brain Coherence

Anyone can generate heart coherence in numerous ways by doing heart-focused meditation, listening to uplifting music or by practicing specific breathing exercises. These can be time consuming in a society that has a busy schedule. Research studies at the HeartMath Institute and elsewhere have shown the easiest and most rapid way to experience coherence is through generating positive emotions (such as appreciation, compassion, love and care) and emotional self-regulation [1].

This new research is starting to show that beyond pleasant subjective feelings, our positive attitudes and emotions exhibit a broad range of objective, interrelated benefits that positively impact our physiological, psychological, and social functioning (Fredrickson, 2002 Isen, 1999). Self-induced, sustained and positive emotions create a sustained shift to a state of system-wide coherence in the body's physiological processes, during which time the coherent pattern of the heartbeat plays a role in facilitating higher cognitive functions.

Evidence also shows that positive emotions affect the body's physiological systems in profound ways. For example positive emotional states speed up the process of recovery of the human cardiovascular system from the after-effects of negative emotions (Fredrickson et al., 2000), alter frontal brain asymmetry (Davidson et al., 2003) and benefit the human body's immune system (Davidson et al. McCraty, Atkinson, Rein, & Watkins, 1996 Rein, Atkinson, & McCraty, 1995). Through regular coherent heart-based practice, it is now possible to use positive emotions to create a shift in the entire physiological system towards a state of global coherence. If kept sustained, this coherence creates vital benefits on numerous levels at the same time, and transforms an individual's life in the process. An example includes increased heart-brain synchronization in a heart that is coherent [2].

This leads to a positive shift in one's perception, behaviour and better cognition (Fredrickson, 2001, 2005 Isen, 1999). This in turn enhances one's faculties such as creativity (Isen, 1998) and intuition (Bolte, Goschke, & Kuhl, 2003). Experiencing positive emotions on a frequent basis has also been shown to predict psychological growth and resilience (Fredrickson, Tugade, Waugh, & Larkin, 2003).

Positive Emotions Lengthen Lifespan

Numerous studies show clear links between positive emotions, longevity and health status (Blakeslee & Grossarth-Maticek, 1996 Danner, Snowdon, & Friesen, 2001 Medalie & Goldbourt, 1976 Moskowitz, 2003 Ostir, Markides, Black, & Goodwin, 2000 Ostir, Markides, Peek, & Goodwin, 2001 Russek & Schwartz, 1997 Seeman & Syme, 1987).

Technology for Inducing Mind/Body Coherence

As described earlier in this book, research has found that utilizing devices that detect when coherence is present (the emWave® meter) combined with emotional self-regulation have proven to be extremely successful for reducing chronic pain, stress, anxiety, anger, fatigue and burnout. These are just a few of the stress-related conditions as there are many more conditions that have been alleviated by practicing self-regulation techniques [3 to 7].

The inner rhythm and state of heart coherence is what many meditation techniques are trying to achieve. The emWave® helps people quickly get into a meditative state and provides feedback. When a person's heart coherence baseline increases, they end up experiencing creativity, vivid mental clarity and enhanced focus throughout their day. This carryover effect allows users to become more intuitive and conscious.

The Coherent Sine Wave

A person's emotions register through various changes in the rhythmic beating patterns of their heart. These patterns change significantly when one experiences different emotions. Hence, negative emotions, such as frustration or anger, show up as erratic, incoherent patterns in the heart's rhythms. Positive emotions, such as appreciation or love show up as a sine wave-like pattern. This denotes coherence in the heart's rhythmic activity (McCraty, Atkinson, Tiller, Rein, & Watkins, 1995 Tiller, McCraty, & Atkinson, 1996). This change in the heart's beating patterns creates changes in the frequency spectra of the electromagnetic fields radiated outward by their heart.

Summary

Sustained positive emotions such as compassion, appreciation and care generate smooth, sine-wave-like patterns in the heart's rhythms, which is a direct reflection of increased order in higher level systems in the human brain, a shift in autonomic balance towards increased parasympathetic activity and increased synchronization between two branches of the ANS. Negative emotions such as anger, frustration, anxiety and worry cause heart rhythm patterns to appear incoherent, which show up as erratic and highly variable. This means less synchronization in the reciprocal action of the sympathetic and parasympathetic branches is taking place in the ANS (McCraty et al., 1995 Tiller et al., 1996). If this de-synchronization in the ANS continues for extended periods of time, it taxes the bodily organs and nervous system.

Lengthened States of Coherence

We all know what it feels like to experience joy and laughter. These are times our heart is strongly coherent without our having to do anything. While the heart goes through periods of natural coherence through experiencing positive emotions, research has found when certain positive emotional states, such as compassion, love or appreciation are intentionally maintained, heart rhythm patterns become coherent that exist for longer periods of time. In physics the explanation of resonance refers to a phenomenon whereby unusually large oscillations are produced in response to an external or internal stimulus whose frequency is the same as, or close to, the natural vibratory frequency of the stationary system. The frequency of vibration produced in this state is defined as the overall resonant frequency of the system which is a natural oscillation that takes place when a person actively feels positive emotions such as compassion, love or appreciation (McCraty et al., 1995). Also this type of resonance also emerges during the stages of deep sleep. The heart is the body's most powerful rhythmic oscillator, which is why it can pull other resonant physiological systems into entrainment with it (Integral Review. December 2009 x Vol.

5, No. 2). (Bradley & Pribram, 1998 Tiller et al., 1996).

Research studies conducted by the HeartMath Institute in Boulder Creek, CA [8] showed that intentional application of these coherence-building techniques, when practiced on a consistent basis, formed a re-patterning / re-writing process in the body's physiological system. This caused the coherence to become familiar to the brain and the nervous system. The final effect being it becomes firmly embedded in the body's neural architecture as a new, stable physiological baseline or set point (McCraty, 2003 McCraty & Childre, 2004 McCraty & Tomasio, 2006). Once this familiar pattern is established, the system strives to maintain this mode automatically. This allows the body to access coherence in a more accessible state during day-to-day activities, and especially so in the midst of challenging or stressful situations.

Coherence Leads to Better Long Term Health

Coherence-building techniques have been shown in numerous studies to generate improvements in key markers of both psychological and physical health, especially the immune system (McCraty et al., 1996 Rein et al., 1995), ANS function and balance (McCraty et al., 1995 Tiller et al., 1996), and the body's DHEA/cortisol ratio (McCraty et al., 1998).

Concerning emotions, significant reductions in anxiety, anger, depression, hostility, burnout and fatigue as well as increased gratitude, peacefulness, caring, contentment and vitality have been reported to take place across diverse populations (Arguelles, McCraty, & Rees, 2003 Barrios Choplin, McCraty, & Cryer, 1997 Luskin et al., 2002 McCraty et al., 1998 McCraty, Atkinson, Lipsenthal, et al. 2003

McCraty, Atkinson, & Tomasino, 2001, 2003).

Significant reductions in blood pressure, glucose and cholesterol levels have also been observed. (McCraty, Atkinson, Lipsenthal, et al., 2003). Also improvements in asthma (Lehrer, Smetankin, & Potapova, 2000) have been noted. Also in individuals with hypertension significant blood pressure reductions have been noted (McCraty, Atkinson, & Tomasino) and reduced depression in patients suffering from congestive heart failure (Luskin et al.) and improved glycemic regulation and improved quality of life in people with diabetes (McCraty, Atkinson, & Lipsenthal, 2000).

Hormones generated by our Heart

Hormones have extremely powerful effects upon our emotions and health. For example the mood swings teens experience is due to an abundance of hormones. The human heart communicates with our mind and body using biochemistry due to the hormones it produces.

The heart was officially classified as a part of the body's hormonal system during 1983. This was confirmed when the newly discovered hormone atrial peptide (atrial natriuretic factor (ANF), atrial natriuretic peptide (ANP) was found to be secreted by the atria of the heart. Further research found that this atrial peptide influenced behaviour and motivation (Telegdy, 1994). Other research found that the heart secretes oxytocin, especially when it is coherent.

Oxytocin. The Bonding Hormone

Henry Dale discovered Oxytocin in 1906 and its molecular structure was completed in 1952 [9]. Oxytocin is known as the love hormone and when it is injected into the cerebrospinal fluid in rats will cause instant penile erections [10].

Oxytocin has been obtained from the posterior lobes of cattle and hog pituitary glands [11]. Oxytocin is produced in the para-ventricular nucleus of the hypothalamus and is released by the posterior pituitary [12]. In women, Oxytocin is released into the blood in response to stretching of the uterus when giving birth. It is also released during stimulation of the nipples when breastfeeding [13]. This helps with the bonding between mother and child.

Also studies have found that the central nervous system releases the cardiac hormone atrial natriuretic peptide (ANP) by the release of oxytocin [14]. Atrial natriuretic peptide is a peptide that allows the heart to maintain a more coherent state.

A reserach study also found that Oxytocin affects Heart Rate Variability [15]. This is an interesting finding because if our coherence is connected to earth's coherence, it shows a type of bond exists between our heart and the earth. This relationship has already been partly confirmed in studies showing stronger geomagnetic activity causes changes in the body, most notably the heart. It would be interesting to see some studies showing the effects of taking Oxytocin during stronger geomagnetic activity. Would it reduce the negative effects of stronger geomagnetic activity? or enhance it due to the bond? I would lean towards the former, because from years of practicing QI Gong type exercises that when geomagnetic activity is stronger the revitalization energy felt after the session is much stronger. This leads me to conclude that QI Gong type exercises during geomagnetic storms produce anti-aging effects, and because QI Gong exercises connect one closer to the earth, and Oxytocin is a bonding peptide, it makes sense that it may enhance one's resistance to stronger geomagnetic storms. Hence this could be why oxytocin exhibits powerful cardioprotective effects [16]. Also because Oxytocin appears to have bonding abilities, and possibly with the energy fields of our earth, it may mean that part of the information received during remote viewing sessions is obtained by tapping into these energy fields that encompass the earth, possibly the energy fields generated by the Schuman resonance and other yet to be discovered energy fields.

Oxytocin has also been found to influence the parasympathic nervous system [17] as well as decrease blood pressure in male mice, but not female mice [18].

As side note, research studies conducted by Dimitrova et al., found that LF and HF bands became reduced during geomagnetic storms [19].

Now that we have a generalized idea that remote viewing involves an association with various energy fields, let's next explore how these fields are interconnected throughout the galaxy, our nervous system and our spinal cord.

References. Chapter 19.

1. Armour, J.A., Neurocardiology--Anatomical and functional principles. 2003, Boulder Creek, CA: HeartMath Research Center, HeartMath Institute, Publication No. 03-011.

2. McCraty et al.: The Coherent Heart. INTEGRAL REVIEW x December 2009 x Vol. 5, No. 221.

3. McCraty, R., Atkinson, M., Tomasino, D., & Bradley, R. T, The coherent heart: Heart-brain interactions, psychophysiological coherence, and the emergence of system-wide order. Integral Review, 2009. 5(2): p. 10-115.

4. Armour, J.A., Potential clinical relevance of the 'little brain' on the mammalian heart. Exp Physiol, 2008. 93(2): p. 165-76.

5. Cantin, M. and J. Genest, The heart as an endocrine gland. Pharmacol Res Commun, 1988.

6. Suppl 3: p. 1-22. 19. Strohle, A., et al., Atrial natriuretic hormone decreases endocrine response to a combined dexamethasone-corticotropinreleasing hormone test. Biol Psychiatry, 1998. 43(5): p. 371-5.

7. Butler, G.C., B.L. Senn, and J.S. Floras, Influence of atrial natriuretic factor on heart rate variability in normal men. Am J Physiol, 1994. 267(2 Pt 2): p. H500-5. 21. Vollmar, A.M., et al., A possible linkage of atrial natriuretic peptide to the immune system. Am J Hypertens, 1990. 3(5 Pt 1): p. 408-11. 18-21

8. Science of the Heart, Volume 2. Exploring the Role of the Heart in Human Performance. An Overview of Research. Rollin McCraty. Feb 2016.

9. The orgasmic history of oxytocin: Love, lust, and labor. Navneet Magon and Sanjay Kalra. Sept 2011.

10. The oxytocin receptor system: structure, function, and regulation. Gimpl G, Fahrenholz F. Physiol Rev. 2001 Apr 81(2):629-83.

11. Studies on high potency oxytocic material from beef posterior pituitary lobes. Pierce Jg, Du Vigneaud V. J Biol Chem. 1950 Sep 186(1):77-84.

12. Gray's Anatomy: The Anatomical Basis of Clinical Practice (41 ed.). Elsevier Health Sciences. 2015. p. 358. ISBN 978-0-7020-6851-5.

13. Chiras DD (2012). Human Biology (7th ed.). Sudbury, MA: Jones & Bartlett Learning. p. 262. ISBN 978-0-7637-8345-7.

14. Oxytocin releases atrial natriuretic peptide by combining with oxytocin receptors in the heart. Gutkowska J. et al. Oct 1997.

15. Oxytocin Increases Heart Rate Variability in Humans at Rest: Implications for Social Approach-Related Motivation and Capacity for Social EngagementAndrew H. Kemp et al. Aug 2012.

16. Oxytocin revisited: its role in cardiovascular regulation. Gutkowska J and Jankowski M. April 2012.

17. Oxytocin specifically enhances valence-dependent parasympathetic responses. Gamer M and Büchel C. Jan 2012.

18. Oxytocin decreases blood pressure in male but not in female spontaneously hypertensive rats. Petersson M. Sept 1997.

19. Dimitrova S, Angelov I, Petrova E. Solar and geomagnetic activity effects on heart rate variability. Natural hazards. 201369:25–37. doi: 10.1007/s11069-013-0686-y.

Chapter 20. Fields of Energy and the Flow of Information

While it can be easy to think that our heart is the only conduit that delivers information to our brain, the fibers in our nervous system act as tiny antennas, perhaps acting as a tuner much you have to turn a dial to receive radio signals.

The Nervous System as an Antenna

It is our theory that the human nervous system acts as a antenna capable of picking up a number of varying biological frequencies which respond to and are tuned into the magnetic fields produced by the hearts of other individuals. The HeartMath Institute has called this **energetic information exchange / energetic communication**. This type of detection ability by the human nervous system is an innate ability which mediates important aspects of true empathy and sensitivity to others along with heightened awareness. The HeartMath Institute states that this energetic communication ability can be enhanced, and in doing so, results in a much deeper level of nonverbal understanding, communication and connection between other people. One key element to this innate ability is its ability to possibly play a role in the therapeutic interaction that takes place between clinicians and their patients. This gives it strong potential to promote rapid   and   lasting   healing.

Research has found that individuals in coherent states are much more sensitive to receiving information that exists in magnetic fields generated by others. This is an important finding because it shows that the information is coming from a magnetic field of a biological nature. This also means that the quantum foam micro-wormholes that exist all around us

[1] may be influencing the heart. Hence the heart may be tuning into thoughts, emotions and information from this region of space or perhaps it may be tuning into information from higher dimensions.

We all go through life learning. Hence life is much like a university, learning from our mistakes, learning to teach others and so on. I outline in my book The Emerald Tablets by Thoth the Atlantean, that the source from which the revitalizing energies come from is information based. Hence information is facilitating the repair of DNA and DNA is acting as a recorder. It is now a fact that DNA can be programmed. As a matter of fact, scientists have actually recorded, than played back a movie inside DNA (Scientists Replay Movie Encoded in DNA. July 2017. NIH.gov).

The Receiving of Information

We exist in a sea filled with information. Recent evidence has shown that a subtle, yet influential "energetic" or electromagnetic communication system is operating at a level just below our conscious level of waking awareness. This information is coupled to energy fields that exist throughout our earth, our solar system and galaxy. Our research performing ARV sessions leads us to conclude that the human nervous system behaves as a sort of antenna, allowing one to tap into the energy contained in these various energy fields, which may also be connected within an intricate web that interacts with other dimensions of time and space.

Research has found that encoded information exists in time intervals that take place between the action potentials in the human nervous system and in patterns in the pulsatile release of human hormones [2]. Research also suggests that the time intervals between a person's heartbeats (HRV) also encode information [2]. This information is than communicated across different varying systems which helps synchronize the whole physiological system.

A direct relationship exists between heart-rhythm patterns and spectral information that is encoded in the frequency field of the magnetic field which extends outwards by the human heart. Hence information about a person's current emotional state becomes encoded in the heart's magnetic field. This information is transferred throughout the body and outwards into their localized environment.

Pulsed Electric Fields

The device we use to enhance the accuracy of our ARV sessions contains a hand operated piezoelectric element that is pressed a number of times to generate a piezoelectric spark. A scientific research study found that exposure of Cronobacter sakazakii or Escherichia coli to high hydrostatic pressure (HHP), pulsed electric fields (PEF) or heat increased the anti-bacterial fighting effect of the citral and its effect to kill bad bacteria (E. coli). The study also found the effects were stronger at pH 4.0 than at pH 7.0. The study concluded limonene exhibited a synergistic effect with heat to help inactivate the pathogen E. coli in fruit juices [3]. It is also interesting to note that pulsed electric fields have been utilized for disinfecting burn wounds [4].

What are Pulsed electric fields (PEF)?

Pulsed electric fields (PEF) consist of a system that generate a high burst of intermittent electrical pulses which cause a disintegration of biological materials with the aim of modifying or destroying the system so as to make it safe to consume. This allows it to be used in the fields of pasteurization, drying and the pelletization of food products.

Piezoelectricity Excites the Nervous System

Our ARV device also includes a power transformer obtained from a used microwave oven as well as rotating magnets utilized from a small D.C. motor. It is a fact that these transformers from microwave ovens have so much wire and iron in them that they naturally generate small amounts of piezoelectricity. It may be that this is having an effect in the surrounding region, affecting the body's parasympathetic nervous system.

Research studies have found that low-intensity ultrasonic waves can remotely excite the central nervous system. A research study found that an ultrasound-induced cavitation of neurons in the central nervous system took place via currents caused by membrane capacitance changes resulting from the ultrasound. The study states that neuronal intra-membrane piezoelectricity influences the nervous system and suggests pressure waves or perturbations that influence the nervous system may be explained due to biological piezoelectric transduction [5].

Piezoelectricity and the Spinal Cord

Parasympathetic fibers that regulate subdiaphragmatic organs travel throughout the body's spinal cord. In this next study we shall see how piezoelectricity is used to rebuild spinal cord neurons.

A research study looked at using piezoelectricity to rebuild spinal cord neurons. The technology utilized oscillating electrical fields delivered to the neurons in rat brains using poly (**vinylidene fluoride**) film. In order to create oscillating electrical fields on the surface of the film, a 50 Hz mechanical vibration was applied. The study found that 4 days later, the neuronal densities had grown by 115% and the neurons grew 79% more neurites, with double the branch points, compared to neurons that were not grown on piezoelectricity stimulated films ($p < 0.001$). The study found that these effects occurred due to the electrical field. The study concluded that oscillating electric fields caused by piezoelectric polymer substrates induce changes in the neurons of the central nervous system leading to significant enhancements in nerve regeneration [6].

Another study found that piezoelectrically active vinylidenefluoride-trifluoroethylene copolymer tubes caused significant enhancement of nerve regeneration compared to controls and that the polarity of the corona poling procedure was key to fabricating piezoelectric materials that play major roles in determining biological responses [7].

Violent Video Games and Cardiac Activity

Coherent states have been connected with states of overall well-being as well as improvements in one's social, cognitive and physical performance. Studies looking at heart-rhythm patterns and their effects upon emotions performed in both natural and laboratory settings and during both intentionally generated and spontaneous emotions have found heart-rhythm patterns change according to a person's emotions [8] [9] [9b].

This effect was explored in a research study looking at people who were playing violent and nonviolent video games. The study discovered that when a person played a violent video game that they displayed LOWER cardiac coherence levels and HIGHER aggression levels. This was in comparison to those who played the nonviolent games [10].

Hence, this is an important discovery. Because if our theory is correct and emotions are not bound by space or time (having a loop feedback type effect), then the heart is picking up emotions from the future which are than interpreted by the brain. Hence during an ARV session when the heart / mind interface is in a coherent state of being, it becomes more sensitive to future emotions meaning if you are remote viewing the future and get the answer wrong, the emotions from the future travel back to the time you are conducting the ARV session, letting you know through feelings, thoughts, sensations and emotions. Hence when one knows what a negative emotion is (failed ARV session, feelings of frustration etc.) and a positive emotion (a successful ARV session) one can interpret the messages / emotions received through the heart which are decoded by the brain.

This corresponds with the study shown earlier at the beginning of this book where people that were shown emotionally charged pictures exhibited stronger pre-stimulis activity compared to those who were shown the non-emotionally charged pictures.

The Amygdala and its role in Emotion

I personally believe that the amygdala plays a key role in ARV sessions. This is due to the fact that the heart's afferent neurological signals affect activity in the brain's amygdala. The amygdala is key to processing emotional information [11]. It also processes what's known as 'emotional memory'. To put it simply, the familiarity of incoming sensory information is processed by the amygdala which uses it to make instantaneous decisions. Because it has a strong relationship to the brain's hypothalamus and autonomic nervous system, at various times it can "hijack" neural pathways that control the autonomic nervous system and elicit an emotional response before our higher brain centers receive any type of sensory information [12]. This is a major finding because it shows a type of intuition is taking place through the experiencing of emotion, rather than the slower type processing of logic and thought. Hence it would imply that emotions are experienced faster than thoughts. Because the processing of emotion may take place faster than the mind can process it, a power stronger than the mind is necessary to alter our perception, override our emotional circuitry, and connect us with our true intuitive feelings. The only organ capable of this is a coherent heart. Hence, in the future the age-old tireless struggle between emotion and intellect won't be resolved by the mind exhibiting its dominance over our emotions, but by creating a coherent balance between our emotional and mental faculties.

As a person gains experience in a certain field, implicit intuitions begin to arise which are derived from their capacity to recognize important environmental cues. They are then able to unconsciously and rapidly match those cues to existing familiar patterns. This allows a person to make an effective diagnosis or solve a perplexing problem with speed an efficiency. Hence, the brain's amygdala may play a major role as the familiarity of incoming sensory information is processed by the amygdala which uses it to make instantaneous decisions.

References. Chapter 20.

1. On the Ground State of Quantum Gravity. S. Cacciatori et al.

2. Heart Rate Variability: New Perspectives on Physiological Mechanisms, Assessment of Self-regulatory Capacity, and Health risk. Rollin McCraty, PhD and Fred Shaffer, PhD, BCB. Jan 2015.

3. Mechanism of bacterial inactivation by (+)-limonene and its potential use in food preservation combined processes. Espina L etl. Feb 2013.

4. Pulsed Electric Fields for Burn Wound Disinfection in a Murine Model. Alexander Golberg et al. Jan 2016.

5. Intramembrane Cavitation as a Predictive Bio-Piezoelectric Mechanism for Ultrasonic Brain Stimulation. Michael Plaksin et al. Jan 2014.

6. Piezoelectric substrates promote neurite growth in rat spinal cord neurons. Royo-Gascon N et al. Jan 2013.

7. Improved nerve regeneration through piezoelectric vinylidenefluoride-trifluoroethylene copolymer guidance channels. Fine EG. Oct 1991.

8. Tiller, W.A., R. McCraty, and M. Atkinson, Cardiac coherence: A new, noninvasive measure of autonomic nervous system order. Alternative Therapies in Health and Medicine, 1996. 2(1): p. 52-65.

9. McCraty, R., et al., The effects of emotions on short-term power spectrum analysis of heart rate variability. American Journal of Cardiology, 1995. 76(14): p. 1089-1093.

9b. Ho, M.-W., The Rainbow and the Worm: The Physics of Organisms2005, Singapore: World Scientific Publishing Co.

10. Violent Video Games Stress People Out and Make Them

More Aggressive. Youssef Hasan. University Pierre Mendès-France, Grenoble, France. Dec 2017.

11. Hopkins, D. and H. Ellenberger, Cardiorespiratory neurons in the mudulla oblongata: Input and output relationhsips, in Neurocardiology, J.A. Armour and J.L. Ardell, Editors. 1994, Oxford University Press: New York. p. 219-244.

12. LeDoux, J., The Emotional Brain: The Mysterious Underpinnings of Emotional Life1996, New York: Simon and Schuster.

Chapter 21. Theories as to How Information from Distant Points in Time is Received

Today we are bombarded with all kinds of information. It is a struggle to know which information is actually of any value. However, we can use our intuition to lead us to what is good information and to do that we need to tap into the informational field within our heart. The heart's field is an important carrier of information. A part of this information may possibly be transmitted / relayed via emotion.

A Theory as to How Information from the Future is Received during ARV Sessions.

Earth's geomagnetic activity affects the heart. Favourable geomagnetic activity levels (Middle Latitude Fredericksburg K-indices between 11 and 4) are favourable to successful ARV sessions. Hence, geomagnetic activity above 11 is creating a type of interference that affects the heart. This interference reduces the signal that comes from the dimension from which time emerges. To put it simply, the future may already exist outside of our dimension. Perhaps mind / heart coherence allows one to tap into this unseen dimension leading to intuition.

Wormholes

Wormholes have such an intriguing air of mystery about them. It kind of reminds one of a worm borrowing through an apple instead of around it, which is actually what takes place when one travels through an actual wormhole as instead of the worm going around the apple it goes through it because space has folded itself. It has been scientifically proven that wormholes exit all around us. These wormholes are so tiny and small that they don't have an effect on our everyday life. These wormholes exist in what is known as the 'quantum foam' as revealed by Kip Thorne (Time Travel and Wormholes: Physicist Kip Thorne's Wildest Theories. Calla Cofield, Space.com. Senior Writer December 19, 2014). It is our hypothesis at the Solar Institute that the nervous system, or a part of it, is receiving information from these wormholes (or possibly higher dimensions) via the quantum foam. This would then mean that the overall information received from the future is coming from one or a combination of the following -

A - An alternate universe

B - Another dimension (a higher dimension that includes time)

C – Through the earth to the remote viewer via a more intense Schuman Resonance

We exist in world filled with invisible fields of all kinds of energy. All cells in our bodies are bathed in an internal and external environment of invisible magnetic forces that are constantly fluctuating [1]. Recent studies are starting to show that fluctuations in these magnetic fields affect the circuits in all biological systems to a lesser or greater degree and that it depends upon the properties of the magnetic fluctuations taking place and the type of biological system [2] [3]. Studies are now confirming that fluctuations in magnetic fields can cause changes in virtually every circuit in living biological systems to a greater or lesser degree. These changes are dependent upon the particular biological system and the properties of the magnetic fluctuations [4] [5] [6]. Thus, global behaviours and human physiological rhythms are synchronized with geomagnetic and solar activity and that disruptions in these fields create visible and emotional effects upon human health and behaviour [7] [8] [9].

There now exist a multitude of studies confirming that sudden / abrupt changes in geomagnetic activity cause increased hospital admissions and show enhanced mortality from strokes, heart attacks and other adverse health effects such as fatigue, mental confusion, depression and traffic accidents. Out of all bodily systems studied, abrupt changes in earth's geomagnetic field have been found to impact the rhythm of the heart the most [10 to 13].

Large scale effects impacting society to greater extents include the crime rate, social unrest, revolutions, increased violence and the frequency of terrorist attacks, all of which have been linked to the solar cycle and the corresponding disturbances in earth's geomagnetic field [14 to 20]. Hence the opposite would be true, **favourable solar weather conditions would strengthen the ability of one to self-regulate their emotions** while remote viewing.

The mechanism for these changes is due to the coupling between the human nervous system resonant frequencies generated by geomagnetic field line resonances as well as the globally propagating magnetic waves called Schumann resonances (SR). These waves take place in the earth-ionosphere resonant cavity. It is now well established in the scientific literature that the ionosphere and earth generate a symphony of resonant frequencies. These frequencies directly overlap with human cardiovascular systems and the human brain. Hence, changes in these resonances influence the functioning of the human autonomic nervous system, cardiovascular system and brain.

Encoded Symbols and the Flow of Information

Signals and messages become encoded and transmitted in physiological systems via the language of patterns (and perhaps even symbology). Just like the heart encodes information, the nervous system contains information within it that is encoded at various time intervals between its patterns of electrical activity and / or action potentials [21]. This type of methodology also applies to humoral communications where biologically relevant information is encoded in the time interval between hormonal pulses [22 to 24]. And future research may find similar effects taking place in the spaces between our thoughts.

It is our hypothesis at the Solar Institute that waves of pressure, most notably high pressure waves, may enhance the sensitivity of information transfer, although research studies are necessary to confirm this.

For example, as we covered earlier in this book, Dr. Yoshiro Nakamats states his greatest ideas occur while taking long underwater swims. He says this is because if the brain has too much oxygen in it, it acts as a deterrent for true inspiration. The trick to starving the brain of oxygen for a short period of time is to dive deep underwater. This will then allow the water pressure to fill the brain with blood. He states that after diving to the bottom of a swimming pool, he holds his breath as long as possible. He calls this the "0.5 seconds before death" zone. While in "the zone" he will visualize an invention and then immediately write the ideas and thoughts he has received on a plexi-glass waterproof tablet. Next he returns to the surface of the swimming pool. The region at the bottom of a swimming pool is a region of high pressure. This matches the region of high barometric air pressure associated with our increased accuracy of remote viewing.

Nerves as a Sensitive Tuning Antenna

Scientific studies have found that a person's alpha rhythm synchronizes to external stimuli. This includes light flashes or sound. However the ability to synchronize to such small subtle electromagnetic signals is a surprising discovery. Hence the electromagnetic signals emanating from quantum foam wormholes may also be collecting information from the collective unconscious via micro wormholes.

Research by the HeartMath Institute found that clear signals were able to be detected from 6 to 18 inches from a person while they were in coherence. Also information that is physiologically relevant is able to be communicated between different individuals at much greater distances and is reflected in synchronized activity. Unlike in most wakeful states of being, synchronization occurring between the heart rhythms of individuals who are close to one another occurs during sleep. In one experiment the information from two subjects that were seated facing each another located 5 feet apart used the Heart Lock-In Technique [25 to 26], which produces sustained states of physiological coherence. The study found that the subjects were able to synchronize themselves to extremely weak external electromagnetic fields (those produced by another person's heart). Perhaps heart coherence between two people is acting much like a tuning fork.

Synchronization in Groups

Firewalling is such a spectacular sport it should be part of the Olympic Games as it exhibits great feats of mental focus and concentration. A research study that involved 38 participants performing a 30-minute Spanish firewalling practice looked at their heart-rate data. The data looked at the synchronized activity between the spectators and the firewalkers. The study discovered fine-grained commonalities of arousal occurred when the fire walking ritual was being performed between the spectators and the firewalkers but not unrelated spectators. The authors in the study concluded that a collective ritual evokes synchronized arousal over time and that this arousal takes place between the active participants, close friends or relatives [27].

Studies now confirm that it is possible for the magnetic field radiating from the human heart to influence the brainwave activity of another person and that this effect can take place at conversational distances. It is the degree of coherence that exists in the receiver's heart rhythms that determines if her/his brainwaves synchronize with the other person's heart rhythm. This means that if a person is in a state of physiological coherence, he or she shows greater sensitivity in registering information patterns and subtle electromagnetic signals that are encoded in the energy fields radiated by another person's heart.

This discovery may lead one to think that they are vulnerable to negative influences occurring from incoherent patterns radiated by the world around them. However upon closer examination, the opposite is in fact true. If a person has developed self-mastery (are able to maintain physiological coherence) their resilience affords them greater internal stability. This in turn creates a type of shielding effect, making them less vulnerable to the negative fields emanating from others in general. Hence it seems increased internal coherence and stability not only allows increased sensitivity, but an opportunity for self-mastery to kick-in if in-coherent fields of energy are encountered.

This is a major finding because it shows that no matter how subtle a signal may be, a coherent state of being is able to not only detect it, but can be immune to its negative effects. Hence a person is much more attuned to another person and is able to better understand and detect deeper meanings behind spoken words. They are able to distinctly sense what a person truly wishes to communicate even if the other person is not exactly clear in what he or she is attempting to say.

Research by Pribram showed low-frequency oscillations generated by the body and heart in the form of afferent neural electrical and hormonal patterns are actually carriers of information in an emotional context. Also higher frequency oscillations discovered in brain EEG activity reflect a person's labelling of feelings and their conscious perception, including their emotions [28].

The HeartMath Institute theorizes that these rhythmic patterns also transmit emotional information via the electromagnetic field into our environment. This emotional transfer of information can be detected by others and is processed in the same manner as internally generated signals. This may explain where 'ghost' hauntings come from. Hence many ghost sightings were environments that experienced emotionally traumatic events.

One example is the commonly reported haunting of movie stages or cemeteries. The sensing of a ghost is really our heart's electromagnetic field picking up on past emotions that occurred a long time ago. These emotions have somehow become 'trapped' in the immediate surrounding region, perhaps in a type of magnetic energy bubble going through a type of time loop. If indeed our earth is enveloped in energy fields, than the emotions from past experiences are trapped in localized regions, perhaps in the form of magnetic fields.

Pressure Waves and the Transfer of Information

Varying intensities of pressure take place around us all the time. They are mostly cycle related going from high to low and back to low to high over a period of days. They are also invisible, making it hard to know which is high or which is low. Our ARV sessions throughout the years were always extremely accurate when the local barometric air pressure had peaked and was just beginning to go into decline.

Our heart generates tiny waves of pressure which travel rapidly through our arteries. These pressure waves have a rate of speed much faster than the flow of blood through our veins. The pressure waves in our arteries push blood cells through the capillaries which provide nutrients and oxygen and to our body's cells and in doing so, expand and contract the arteries. This causes them to generate a large electrical voltage (large in proportion to its size). The pressure these waves produce also create pressure at the cells in a rhythmic manner which causes some of the **proteins to generate electrical current**, which comes from the "squeezing motion" taking place in the arteries. Hence our arteries are like wires, allowing bioelectrical currents to flow through them.

Research by the HeartMath Institute found that changes in the brain's electrical activity take place when the wave of blood-pressure reaches the brain approximately 240 milliseconds after systole [29].

During physiological coherence, the body's physiological functions are more efficient and radiate stronger coherent electromagnetic fields into the environment [30]. Hence a region of space that is coherent may become phase locked with a heart that is coherent, contributing to an exchange of information. Another theory is the information comes from the higher self which is able to look out into the collective consciousness. Because our ARV sessions are conducted at midnight and as we covered earlier the body / mind is more coherent when asleep, a similar phase lock type effect may be occurring between the collective unconscious and the person who is in a coherent state of being.

Now that we have a complete understanding of how self-regulation can lead to self-mastery, let's explore its effects on health in the next chapter.

References. Chapter 21.

1. Halberg, F., et al., Cross-spectrally coherent ~10.5- and 21-year biological and physical cycles, magnetic storms and myocardial infarctions. Neuroendocrinology, 2000. 21: p. 233-258.

2. McCraty, R., Atkinson, M., Tomasino, D., & Bradley, R. T, The coherent heart: Heart-brain interactions, physiological coherence, and the emergence of system-wide order. Integral Review, 2009. 5(2): p. 10-115.

3. Halberg, F., et al., Cross-spectrally coherent ~10.5- and 21-year biological and physical cycles, magnetic storms and myocardial infarctions. Neuroendocrinology, 2000. 21: p. 233-258.

4. Bergiannaki, J.-D. Paparrigopoulos, T.J. Stefanis, C.N. Seasonal pattern of melatonin excretion in humans: Relationship to day length variation rate and geomagnetic field fluctuations. Experientia 1996, 52, 253–258.

5. Hamer,J.R.Biological Entrainment of the Human Brain by Low Frequency Radiation. Northrop Space Laboratories: Hawthorne, CA, USA, 1965 pp. 65–199.

6. Caswell, J.M. Carniello, T.N. Murugan, N.J. Annual incidence of mortality related to hypertensive disease in Canada and associations with heliophysical parameters. Int. J. Biometeorol. 2016, 60, 9–20.

7. Lean, J. Evolution of the sun's spectral irradiance since the maunder minimum. Geophys. Res. Lett. 2000, 27, 2425–2428.

8. Tapping, K. Recent solar radio astronomy at centimeter wavelengths: Thetemporalvariabilityofthe10.7-cm flux. J. Geophys. Res. Atmos. 1987, 92, 829–838.

9. Babayev, E. Crosby, N. Obridko, V. Rycroft, M. Potential

effects of solar and geomagnetic variability on terrestrial biological systems. In Advances in Solar and Solar Terrsetrial Physics Maris, G., Demetrescu, C., Eds. Research Signpost: Kerala, India, 2012 pp. 329–376.

10. J.B. Reif, J.S. Yost, M.G. Geomagnetic disturbances are associated with reduced nocturnal excretion of a melatonin metabolite in humans. Neurosci. Lett. 1999, 266, 209–212.

11. Bergiannaki, J.-D. Paparrigopoulos, T.J. Stefanis, C.N. Seasonal pattern of melatonin excretion in humans: Relationship to day length variation rate and geomagnetic field fluctuations. Experientia 1996, 52, 253–258.

12. Malin, S.R.C. Srivastava, B.J. Correlation between heart attacks and magnetic activity. Nature 1979, 277, 646–648.

13. Giannaropoulou, E. Papailiou, M. Mavromichalaki, H. Gigolashvili, M. Tvildiani, L. Janashia, K. Preka-Papadema, P. Papadima, T. A study on the various types of arrhythmias in relation to the polarity reversal of the solar magnetic filld. Nat. Hazards 2014, 70, 1575–1587.

14. Cornélissen, G. Halberg, F. Breus, T. Syutkina, E.V. Baevsky, R. Weydahl, A. Watanabe, Y. Otsuka, K. Siegelova, J. Fiser, B. Non-photic solar associations of heart rate variability and myocardial infarction. J. Atmos. Sol. Terr. Phys. 2002, 64, 707–720.

15. Villoresi,G. Ptitsyna,N.G. Tiasto,M.I. Iucci,N. Myocardial infarct and geomagnetic disturbances: Analysis of data on morbidity and mortality. Biozika 1998, 43, 623–632. (In Russian).

16. Malin, S.R.C. Srivastava, B.J. Correlation between heart attacks and magnetic activity. Nature 1979, 277, 646–648.

17. Stoupel,E. Sudden cardiac deaths and ventricular extrasystoles on days of four levels of geomagnetic activity. J.

Basic Physiol. Pharmacol. 1993, 4, 357–366.

18. Persinger, M.A. Sudden unexpected death in epileptics following sudden, intense, increases in geomagnetic activity: Prevalence of effect and potential mechanisms. Int. J. Biometeorol. 1995, 38, 180–187.

19. Knox,E.G. Armstrong,E. Lancashire,R. Wall,M. Hayes,R.Heartattacksandgeomagneticactivity. Nature 1979, 281, 564–565.

20. Giannaropoulou, E. Papailiou, M. Mavromichalaki, H. Gigolashvili, M. Tvildiani, L. Janashia, K. Preka-Papadema, P. Papadima, T. A study on the various types of arrhythmias in relation to the polarity reversal of the solar magnetic field. Nat. Hazards 2014, 70, 1575–1587.

21. Pribram, K.H., Brain and Perception: Holonomy and Structure in Figural Processing1991, Hillsdale, NJ: Lawrence Erlbaum Associates, Publishers.

22. Prank, K., et al., Coding of time-varying hormonal signals in intracellular calcium spike trains. Pac Symp Biocomput, 1998: p. 633-44.

23. Schofl, C., K. Prank, and G. Brabant, Pulsatile hormone secretion for control of target organs. Wiener Medizinische Wochenschrift, 1995. 145(17-18): p. 431-435.

24. Schofl, C., et al., Frequency and amplitude enhancement of calcium transients by cyclic AMP in hepatocytes. Biochem J, 1991. 273(Pt 3): p. 799-802.

25. Childre, D. and H. Martin, The HeartMath Solution 1999, San Francisco: Harper. SanFrancisco.

26. McCraty, R., et al., The impact of a new emotional self-management program on stress, emotions, heart rate variability, DHEA and cortisol. Integr Physiol Behav Sci, 1998. 33(2): p. 151-70.

27. Synchronized arousal between performers and related spectators in a fire-walking ritual. Ivana Konvalinka et al. May 2011.

28. Pribram, K.H. and F.T. Melges, Psychophysiological basis of emotion, in Handbook of Clinical Neurology, P.J. Vinken and G.W. Bruyn, Editors. 1969, North-Holland Publishing Company: Amsterdam. p. 316-341.

29. Science of the Heart, Volume 2. Exploring the Role of the Heart in Human Performance. An Overview of Research. Rollin McCraty. Feb 2016.   Energetic   Communication. Chapter 6.

30. Tiller, W.A., R. McCraty, and M. Atkinson, Cardiac coherence: A new, noninvasive measure of autonomic nervous system order. Alternative Therapies in Health and Medicine, 1996. 2(1): p. 52-65.

Chapter 22. Techniques for Better Health via Self-Mastery

By reading about the latest information on coherence in this chapter, it can save considerable amounts of money on health bills and emotional trauma. Good health is not just a blessing, but needs care and recognition for its ability to make our lives less   stressful.

It is now an established scientific fact that an estimated 60% to 80% of doctor visits are related to stress [1 to 3]. Research studies show usage of self-regulation techniques increases one's parasympathetic activity (HF power) [4]. This in turn causes increases in DHEA and significant reductions in cortisol over a 30-day period [5]. Other studies show significantly lowered blood pressure and stress measures.

Self-Regulation Reduces dependence upon Pharmaceuticals
Many of us may think that we need a doctor's permission or special herbal formulas to enjoy good health and long life. In some cases that may be true. But what if coherence could improve   an   existing   negative   health   condition?

Employees that were diagnosed with hypertension that volunteered to enroll in a workplace-based risk-reduction program showed significant reductions in their blood pressure compared to the control group after they had used HeartMath self-regulation techniques for three months. And what is surprising is one of the participants was able **to completely discontinue his antihypertensive medication usage** completely following completion of the research study [6].

This is a major and significant finding because it shows that specific **self-regulation techniques can positively affect the biochemistry of an individual** and alter it in a beneficial way.   This may also mean that certain self-regulation techniques may be as powerful or more powerful than some medications or herbal formulas and that they may show a synergistic effect with specific herbs or medications and possibly during lunar phases. Further research is needed to confirm this hypothesis.

Another study looked at people who took hypertensive medications. They were taught the Quick Coherence Self-Regulation Technique and were also given a heart rate variability (HRV) coherence-training device [7]. This is a device that lets your body know if and when you are in coherence.   One example is the emWave®, which I have successfully used for years. Anyway, back to the study. The study found that the greatest reductions in the participant's blood pressure occurred when they took their medications and also practiced the Quick Coherence Self Regulation Technique using the HRV coherence electronic device [8]. Another interesting finding of the study was that the group not taking medications showed greater reductions in their blood pressure compared to the medications/relaxation group [8].

Coherence improves Cognitive Abilities

Anti-aging research is starting to discover that the more a person retains healthy cognitive abilities, that the longer they live. Wouldn't it be great to know a secret method that one could use that would reduce their dependence upon herbs or medications that keeps their mental faculties sharp well into old   age?

Studies have found that increased levels of heart-rhythm coherence were associated with major improvements in a

person's cognitive performance [9 to 11]. This is a major finding. This is because cognitive decline is one of the major factors related to aging. Also research at the Solar Institute over the years has discovered that many of the major herbs that combat the aging process, that the very best ones happen to be neuro-protective herbs. To put it simply, when one can keep their brain strong and healthy (strengthened neuro-cognition), one may be able to live a longer lifespan.

Other reserach studies found major outcomes in **reaction-time** and discrimination-type experiments where complex domains of cognitive function, such as academic performance and memory are improved via self-regulation techniques [12 to 13]. Healthier emotional and cognitive functioning results in significant reductions in stress, anger, hostility, depression, anxiety, burnout and fatigue. It also increases gratitude, peacefulness, caring, contentment, resilience and vitality. This effect has been noted across diverse populations [14 to 19].

References. Chapter 22.

1. Nerurkar, A., et al., When physicians counsel about stress: results of a national study. JAMA Intern Med, 2013. 173(1): p. 76-7.

2. Avey, H., et al., Health care providers' training, perceptions, and practices regarding stress and health outcomes. J Natl Med Assoc, 2003. 95(9): p. 833, 836-45.

3. Cummings, N.A. and G.R. Vanden Bos, The twenty years Kaiser-Permanente experience with psychotherapy and medical utilization: implications for national health policy and national health insurance. Health Policy Q, 1981. 1(2): p. 159-75.

4. Tiller, W.A., R. McCraty, and M. Atkinson, Cardiac coherence: a new, noninvasive measure of autonomic nervous system order. Altern Ther Health Med, 1996. 2(1): p. 52-65.

5. McCraty, R., et al., The impact of a new emotional self-management program on stress, emotions, heart rate variability, DHEA and cortisol. Integr Physiol Behav Sci, 1998. 33(2): p. 151-70.

6. Science of the Heart, Volume 2. Exploring the Role of the Heart in Human Performance. An Overview of Research. Rollin McCraty. Feb 2016. Health Outcome Studies. Chapter 8. Page 54.

7. McCraty, R., M. Atkinson, and D. Tomasino, Impact of a workplace stress reduction program on blood pressure and emotional health in hypertensive employees. J Altern Complement Med, 2003. 9(3): p. 355-69.

8. Alabdulgader, A., Coherence: A Novel Nonpharmacological Modality for Lowering Blood Pressure in Hypertensive Patients. Global Advances in Health and

Medicne, 2012. 1(2): p. 54-62.

9. McCraty, R., Atkinson, M., Tomasino, D., & Bradley, R. T, The coherent heart: Heart-brain interactions, psychophysiological coherence, and the emergence of system-wide order. Integral Review, 2009. 5(2): p. 10-115.

10. Lloyd, A., Brett, D., Wesnes, K., Coherence Training Improves Cognitive Functions and Behaviour In Children with ADHD. Alternative Therapies in Health and Medicine, 2010. 16(4): p. 34-42.

11. Ginsberg, J.P., Berry, M.E., Powell, D.A., Cardiac Coherence and PTSD in Combat Veterans. Alternative Therapies in Health and Medicine, 2010. 16(4): p. 52-60.

12. McCraty, R., Atkinson, M., Tomasino, D., & Bradley, R. T, The coherent heart: Heart-brain interactions, psychophysiological coherence, and the emergence of system-wide order. Integral Review, 2009. 5(2): p. 10-115.

13. Bradley, R.T., McCraty, R., Atkinson, M., Tomasino., D., Emotion Self-Regulation, Psychophysiological Coherence, and Test Anxiety: Results from an Experiment Using Electrophysiological Measures. Applied Psychophysiology and Biofeedback, 2010. 35(4): p. 261-283.

14. Luskin, F., et al., A controlled pilot study of stress management training of elderly patients with congestive heart failure. Preventive Cardiology, 2002. 5(4): p. 168-172, 176.

15. Arguelles, L., R. McCraty, and R.A. Rees, The heart in holistic education. Encounter: Education for Meaning and Social Justice, 2003. 16(3): p. 13-21.

16. Barrios-Choplin, B., R. McCraty, and B. Cryer, An inner quality approach to reducing stress and improving physical and emotional wellbeing at work. Stress Medicine, 1997.

13(3): p. 193-201.

17. McCraty, R., Heart-brain neurodynamics: The making of emotions. 2003, Boulder Creek, CA: HeartMath Research Center, HeartMath Institute, Publication No. 03-015.

19. McCraty, R. and M. Atkinson, Spontaneous heart rhythm coherence in individuals practiced in positive-emotion-focused techniques. Unpublished data, 1998. 280. McCraty, R., et al., Impact of the Power to Change Performance program on stress and health risks in correctional officers. 2003: Boulder Creek, CA: HeartMath Research Center, HeartMath Institute, Report No. 03-014, November 2003.

Chapter 23. Self-Mastery Skills for Excelling in Mathematics. Practical Applications for Engineers and Physicists.

It is interesting to note that HeartMath coherence improves the ability for one to tackle math problems, and the name of the various coherence self-regulation techniques described in this book are termed "Heart-MATH exercises". The sole purpose of practicing Heart coherence is to eliminate anxiety, which for some people just hearing the word math invokes feelings   of   anxiety. The key to mastering mathematics is reduced anxiety.

A study found that inhaling the substance linalool greatly increased the mathematical accuracy of people performing mathematical   calculations.

Reference:
The Effects of Linalool and Peppermint Aroma on Cognitive Performance. Kaufman, Robert et al. May 2017. The Ohio State University. Department of Psychology Undergraduate Research Theses 2017.

High amounts of Linalool have been found in Neroli and Petitgrain essential oils (Singlet oxygen scavenging activity and cytotoxicity of essential oils from rutaceae. Ao Y et al. July 2006). Hence these would be great essential oils to have around if you worked in an engineering or physics environment. Linalool has also been shown to reduce anxiety. Linalool has also been found in chocolate cacao (Differentiation of chocolates according to the cocoa's geographical origin using chemometrics. Cambrai A et al. Feb 2010). A research study found that when mice inhaled linalool that it reduced their anxiety and when inhaled at high doses it impaired their memory.

Reference

Effects of inhaled Linalool in anxiety, social interaction and aggressive behavior in mice. Linck VM et al. July 2010.

Lavendar essential oil contains up to 37.4% linalool.

Reference

Exploring Pharmacological Mechanisms of Lavender (Lavandula angustifolia) Essential Oil on Central Nervous System TargetsVíctor López,. et al. May 2017.

For some people just saying the word 'math' generates feelings of anxiety, not to mention past experiences one might have had trying to solve what may seem to some "slippery" and hard to grasp equations. Self-regulation coherence techniques seem especially suited to people who practice physics and math, perhaps due to the fact that math can create anxiety for some people.

A research study involving Heart-rhythm coherence feedback that used the FreezeFramer (now called emWave Pro) involved studies working on math problems. While they worked on their math problems they activated the emWave Pro. During these sessions, students could observe their reactions to extremely difficult mathematical calculations which was reflected in the HRV feedback. They practiced self-regulation coherence techniques at the same time while working on their math problems. This allowed them to gain valuable insight into their emotional responses concerning self-regulation. This also led to an increase in

their intuition which helped find ways to solve their math problems. The study found that the students became very responsive to the program. Over the course of a number of years their math results improved and the teachers found new ways to integrate the use of self-regulation techniques in order for students to become proficient at math [1].

Summary
During year one when the teachers integrated self-regulation techniques, an increase of 19% in math scores occurred. By year three, this increased to 24%, utilizing Compass Test Scores. This was compared to the classes that did not use the self-regulation techniques [1]. These were powerful gains considering the program was of short duration and the key purpose was emotion-management skills rather than on formal math instruction. Hence this shows that calm and level emotions have a profound effect on school studies, most notably subjects that may evoke feelings of anxiety for some people.

By year four self-regulation techniques and HeartMath techniques were completely integrated into the curriculum. Year four produced the very best results out of all years, far exceeding the teacher's expectations.

Results
A significant ($p < 0.001$) improvement averaging an increase of 73% in math scores.

The program than expanded into a college prep program with the intention of completely eliminating math remediation classes. The program was integrated into 11th year math classes at a local high school. The study involved

16 students and a math teacher who utilized self-regulation techniques via the HeartMath System. In the classroom were Four Freeze-Framer Stations where students could sit and practice self-regulation techniques throughout the class period. The college prep program was also a success [2].

This is a particularly unique study in that it shows self-regulation techniques may be of value to engineers or physicists working on particularly difficult problems. Speaking from personal experience, when I have had an engineering challenge, I have found that taking a few drops of a Banisteriopsis caapi tincture / extract before bed allows my mind to sleep on the problem. The following morning the engineering problem is solved. This could be due to the fact that Banisteriopsis caapi induces extremely deep relaxation in the body and possibly the heart region which allows information that solves problems to more easily flow. Hence, future studies may find that Banisteriopsis caapi positively affects heart rate variability (HRV).

The studies conducted by Drs. Vislocky and Leslie were the first of their kind integrating self-regulation techniques into a mathematics learning environment. Their studies exhibited strong evidence that integration of coherence-building technologies and tools into an instructional environment is an effective way to enhance student academic performance and learning, as well as better prepare students for entry into higher education.

Law Enforcement and Self-Regulation Techniques

Another study looked at the degree and nature and of physiological activation experienced by police officers on the job who utilized HeartMath's Resilience Advantage Training

Program.

This program involved police officers from Santa Clara County, California [3]. The assessment included particulars such as stress coping and interpersonal skills, work performance, vitality, workplace effectiveness and social climate, emotional well-being, physiological recalibration and family relationships all following acute stressors in their daily lives. The study found that resilience-building techniques improved the police officers' ability to self-regulate and recognize their responses to stressors in both personal and work contexts. The self-regulation techniques showed that police officers experienced major reductions in negative emotions, stress and depression.

Key Benefits of the study found the following:

Enhanced awareness and self-management of stress reactions.

Higher confidence, clarity and balance under acute stress.

Psychological and physiological recalibration following acute stress.

Better work performance.

Improved communication, reduced competition and greater cooperation amongst work teams.

Reduced sadness distress, anger and fatigue.

Reduced physical stress symptoms and sleeplessness.

Enhanced vitality and peacefulness.

Better relationships with family and listening.

References. Chapter 23.

1. Science of the Heart, Volume 2. Exploring the Role of the Heart in Human Performance. An Overview of Research. Rollin McCraty. Feb 2016. Outcome Studies in Education. Chapter 9. Page 79.

2. Science of the Heart, Volume 2. Exploring the Role of the Heart in Human Performance. An Overview of Research. Rollin McCraty. Feb 2016. Outcome Studies in Edicaiton. Chapter 9. Page 80.

3. McCraty, R. and M. Atkinson, Resilence Training Program Reduces Physiological and Psychological Stress in Police Officers. Global Advances in Health and Medicne, 2012. 1(5): p. 44-66.

Chapter 24. The Planetary-wide Global Coherence Experiment

The cells in our body are constantly bathed in internal and external fluctuating invisible magnetic fields which affect virtually all our cells, their associated circuits and related biological systems [1]. Hence, numerous physiological rhythms that occur in humans and global collective behaviours are synchronized with geomagnetic and solar activity. Disruptions in these fields create adverse effects on human behaviour and health. Coherent states of being allow one to become more connected to the external fields of solar and cosmic activity as well as allowing one a deeper connection to human emotion and consciousness. This new awareness than affects the emotional and mental states of others' consciousness. Hence it broadens our view of how interconnectedness can be intentionally utilized to better shape the future of the world we live in.

The cognitive functions, emotions and behaviour health of humans, including animals, are affected by planetary energetic and magnetic fields. Earth's magnetic field acts as a carrier of biologically relevant information that connects all living things. Individuals affect this global information field with the effects being more sustained if they involve strong emotions. A large group of people creating heart-centered states of love, care and compassion generate stronger coherent fields that help offset the current planetary wide discord [2].

The above data also includes a related hypothesis which states human consciousness and emotions interact with and encode information in the geomagnetic field of the earth.

Hence, information is communicated at a nonlocal level between people at a subconscious level. This causes living systems to be linked to one another which in turn influences the overall collective consciousness. Hence we see a feedback loop occurring between human beings and earth's energetic systems. As coherently aligned individuals intentionally create physiologically coherent waves, a more effective resonation takes place which in turn creates a stronger imprint in the encoded information taking place in earth's magnetic fields.

Summary
Magnetic fields act as carrier waves positively influencing all living systems that are contained within the field environment, including the collective consciousness [3].

The Global Consciousness Project (GCP)
Mankind is just starting to discover that we are all connected to one another and that this connection expresses itself through coherence. Hence, it may be possible to create a coherent global standing wave form. We have seen the opposite to be true, where earth's magnometer showed extreme disturbances during the 911 attacks as well as other events where large numbers of people were affected [3b].

The HeartMath Institute in Boulder Creek Colorado, has designed and implemented what is known as The Global Coherence Monitoring System (GCMS). This system is designed to gather scientific data that occurs in Earth's electromagnetic fields [4]. It consists of state-of-the-art magnetometers which are located in specific locations on earth. Each GCMS site contains ultrasensitive magnetic field detectors that measure magnetic resonances occurring in earth's ionosphere cavity. These are resonances which are generated by subtle vibrations that occur in earth's geomagnetic field lines. It also measures ultra-low frequencies which take place in earth's magnetic fields. Variations in all these have been shown to affect human emotional processes, health and mental behaviour. The activity can be viewed in real time on a spectrograph which is shown below.

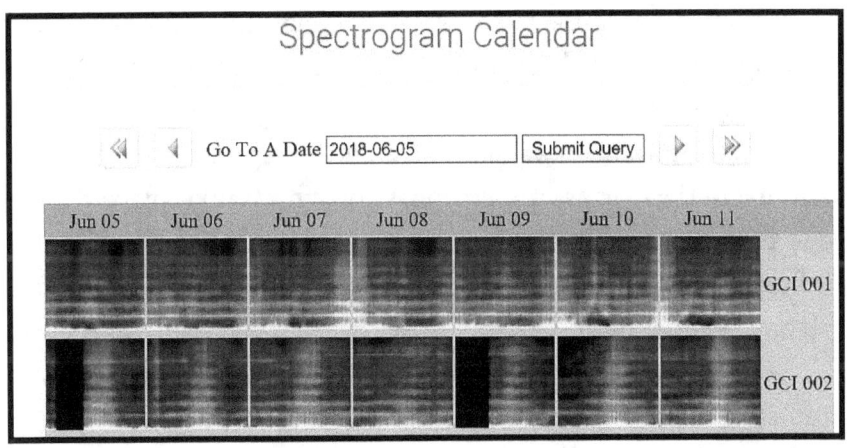

A special part of this project includes seeing if changes in earth's magnetic fields occur before natural catastrophes take place such as volcanic eruptions, earthquakes or human events such as terrorist attacks and social unrest.

When an event is characterized by widespread and deep compassion, the GCP (Global Consciousness Project) effects become stronger [5]. It is already a fact that peaks in geomagnetic activity occur much more often during a major terrorist attack, which we go into greater detail in the book Improve your Remote Viewing Accuracy Techniques using Quantum Microtubules. Also recent studies show that changes occur in earth's ionosphere before an earthquake strikes [6] [7]. As we covered earlier on in this book, Schumann resonances are generated by lightning in the cavity formed by the Earth's surface and its ionosphere.

Melatonin and Geomagnetic Storms

Research studies conducted by Burch et al. [8] and Rapoport et al. [9] found a drop in melatonin levels would take place when there was an increase in geomagnetic and solar activity. This is important because it shows that melatonin may be linked to aging. It is a well established fact that melatonin declines as one ages. And because geomagnetic storms cause a decrease in melatonin, the body may age more rapidly during major geomagnetic storms, the strongest of which take place during the Spring and Fall Equinoxes [10]. Hence a sort of anti-aging serum may be able to be developed that one takes when strong geomagnetic storms occur. It is also interesting to note that the apple is associated with long life and apples contain melatonin [11].

Acute heart disease, heart attacks, accelerated aging, cancer, neurological disorders, as well as other diseases affect the body's melatonin levels. And as we have shown extensively throughout this book, clinical studies have found significant changes occur in a person's blood pressure, aggregation and coagulation, blood flow, cardiac arrhythmia and heart rate variability (HRV) when geomagnetic activity is at above or below average levels. All of these conditions are also influenced by melatonin levels [12] [13].

Melanin as a Semiconductor

Both melanin and melatonin are made from amino acids, and both have a relationship to light. In electronic semiconductors, such as the ones used in cell phones and computers, the electrons transport the electrical current. However our brains and muscles, which are biologically based, the ions transport the current.

A research study found that melanin transports both electrons and ions. This allows for an interface to occur between conventional electronics and biological systems using ion-and-electron conducting biomaterials such as melanin. This is because melanin can 'talk' to both ionic and electronic control circuitry allowing a strong connection.

Reference
Role of semiconductivity and ion transport in the electrical conduction of melanin. Albertus B et al. May 2012.

Summary
Melanin is affected by electromagnetic forms of energy. This may mean that melanin may also exhibit resonance effects, although further research is necessary to validate this.

References. Chapter 24.

1. Halberg, F., et al., Cross-spectrally coherent ~10.5- and 21-year biological and physical cycles, magnetic storms and myocardial infarctions. Neuroendocrinology, 2000. 21: p. 233-258.

2. Science of the Heart, Volume 2. Exploring the Role of the Heart in Human Performance. An Overview of Research. Rollin McCraty. Feb 2016. Global Coherence Research: Human-Earth Interconnectivity. Chapter 11. Page 89.

3. McCraty, R., A. Deyhle, and D. Childre, The global coherence initiative: creating a coherent planetary standing wave. Glob Adv Health Med, 2012. 1(1): p. 64-77.

3b. Figure 10. The Global Coherence Initiative: Creating a Coherent Planetary Standing Wave. Rollin McCraty. et al. March 2012.

4. Science of the Heart, Volume 2. Exploring the Role of the Heart in Human Performance. An Overview of Research. Rollin McCraty. Feb 2016. Global Coherence Research: Human-Earth Interconectivity.Chapter 11. Page 90.

5. Nelson, R. Scientific Evidence for the Existence of a True Noosphere: Foundation for a Noo-Constitution. in World Forum of Spiritual Culture. 2010. Astana, Kazakhstan.

6. Uyeda, S., et al., Geoelectric potential changes: possible precursors to earthquakes in Japan. Proc Natl Acad Sci U S A, 2000. 97(9): p. 4561-6.

7. Kopytenko, Yu A., et al. "Detection of ultra-low-frequency emissions connected with the Spitak earthquake and its aftershock activity, based on geomagnetic pulsations data at Dusheti and Vardzia observatories." Physics of the Earth and Planetary Interiors 77.1(1993): p. 85-95

8. Burch, J.B., Reif, J.S., Yost, M.G. Geomagnetic disturbances are associated with reduced nocturnal excretion of a melatonin metabolite in humans. Neuroscience Letters, 1999. 266: p. 209-212.

9. Rapoport, S.I., Blodypakova, T.D., Malinovskaia, N.K., Oraevskii, V.N., Meshcheriakova, S.A., Breus, T.K. and Sosnovskii, A.M., , Magnetic storms as a stress factor. Biofizika, 1998. 43(4): p. 632-639.

10. Biotropic effects of geomagnetic storms and their seasonal variations. Kuleshova VP et al. Sept 2001.

11. Melatonin in Apples and Juice: Inhibition of Browning and Microorganism Growth in Apple Juice. Zhang H et al. Feb 2018.

12. Halberg, F., et al., Time Structures (Chronomes) of the Blood Circulation, Populations' Health, Human Affairs and Space Weather. World Heart Journal, 2011. 3(1): p. 1-40.

13. Doronin, V.N., Parfentev, V.A., Tleulin, S.Zh, .Namvar, R.A., Somsikov, V.M., Drobzhev, V.I. and Chemeris, A.V., Effect of variations of the geomagnetic field and solar activity on human physiological indicators. Biofizika, 1998. 43(4): p. 647-653.

Chapter 25. The Collective Subconscious

Remote Viewing, Coherence and Environmental Variables

This next section is a key part to successful remote viewing, as it offers solid scientific evidence that specific solar and terrestrial conditions are favourable to successful associative remote viewing sessions. Just as it took a few years for the scientific studies to appear confirming my hypothesis that solar weather influences health and behaviour, the same is now taking place showing that these same solar influences affect one's ability to successfully remote view the dow jones or   FOREX.

Ease of Being and State of Mind. The power of the Collective Unconscious

As stated earlier, one hypothesis of where the subconscious mind gets its information during associative remote viewing sessions is by connecting with the overall global collective unconscious and gathering the required information. This information may involve future intentions to buy or sell particular stocks or currencies. Also because the mind is more coherent when one is asleep, our ARV sessions are conducted at around midnight, when connection with the collective unconscious is much easier due to less anxiety which causes interference.

An interesting finding came from our ARV sessions which took place looking at the future FOREX currencies. All our ARV sessions looking at the British Pound against the Yen always would fail. We have concluded that the reason this was, was due to the fact that when the ARV session was conducted, the British FOREX markets were open. Hence a type of interference may have been taking place. As per our ARV sessions looking at the Dow Jones or USD and AUD markets, these markets were closed when our ARV sessions were conducted, and we had much, much better success remote viewing their future positions. This was when the ARV sessions were conducted at mid-night in Hawaii. Hence, this would imply when the markets are open, fear and anxiety is present, making it hard to remote view that specific target. When the markets are closed, this anxiety is not as strong. We can think of this in terms of experiencing weather. If it has been raining for weeks and the next few days are filled with sunshine, people in that community are more up-beat and happy. Hence the overall collective consciousness in that region is positive. As a matter of fact the 'window' of favourable solar activity that takes place each ARV session is a period where overall no major terrorist attacks have ever happened and there exists a stronger overall global harmony. You can read the report published showing this in greater detail in my book Solar Flares and their Effects upon Human Behaviour and Health.

Solar Weather's Effect on Mood and Anxiety

A research study looked at solar wind speeds and discovered that when it increased and earth's geomagnetic field was disturbed, that levels of anxiety, fatigue and mental confusion in the participants increased [1]. This is a major finding because the majority of our successful ARV sessions took place when the solar wind speed was at favourable levels (low levels at around 350). Also successful ARV sessions would take place when the Fredericksburg K-indices were between 11 and 7.

http://legacy-www.swpc.noaa.gov/ftpdir/indices/DGD.txt

The aforementioned study also discovered some surprising findings. The sun's 10.7 cm solar radio flux index was positively correlated with improved positive affect and reduced fatigue [1]. Hence less fatigue means more mental strength, focus and concentration. Our ARV sessions were always extremely accurate when the sun's 10.7cm radio flux had been increasing the previous 24 hours and more-so if it had been increasing for 3 consecutive days.

Increased Polar Cap Activity is Associated with enhanced Feelings of Anxiety and Irritability

The study also found that when polar cap activity increased, overall feelings of well being among the participants decreased. Hence stronger polar cap activity may also be beneficial to successful ARV sessions, although further research is needed to confirm this.

What Does Polar Cap Activity Mean?

The polar geomagnetic activity comes from the interaction between the interaction of the sun's solar wind and earth's magnetosphere. These interactions are given values and called Polar Cap Activity. The values are derived from polar magnetic variations. Solar wind effects are calculated via the geo-effective electric field (EM) and the dynamical pressure (PDYN). In summary it is a display of the saturation-like effects of earth's magnetosphere and the solar wind. When solar wind is at low speeds, there is very little impact or no impact on the PC index. Sharp increases in dynamical pressure generate variations in the Polar Cap index. Hence it is a good indicator of sudden changes and how extreme changes in the solar wind are.

Reference
The Polar Cap (PC) indices: Relations to solar wind parameters and global magnetic activity. P. Stauning et al. Dec 2008.

General Summary
Data from the HRV report and the Interconnectedness Study show when earth's magnetic field is calmer or the sun's F10.7cm solar radio flux increases, participants felt better, were more emotionally and mentally stable and had HIGHER levels of HRV. Alternatively when earth's magnetic field becomes disturbed, a person's mental clarity and emotional well-being become adversely affected [1]. This is a major finding that can be easily overlooked in all the raw data.

To summarize, when a person practices self-regulation techniques, such as the quick coherence exercise for example, while solar weather conditions are favourable, a coherent synergy forms between the earth (quite possibly ionosphere) and the person who is in coherence. This greatly amplifies the effects of where the intention behind this coherence is directed, be it healing, the ability to solve complex math problems, to relieve stress and tension or to remote view. It is a fact that during favourable solar conditions, which I call a condition green period (www.ez3dbiz.com/in_depth.html) that it is the very best time to get a massage as more stress and tension is released from the body during this phase of solar activity than from any other cycle. Also practicing self-regulation techniques while solar weather conditions are favorable, 4 to 5 days before a geomagnetic storm may provide a barrier or shield against the negative effects from future geomagnetic storms that negatively impact health and well-being. Further research is necessary to confirm this.

Low HRV Levels are Good for Health

HRV levels that are low, when adjusted for age, is a strong and independent predictor of future health problems. This is the case in healthy people and also in people suffering from coronary artery disease [2] [3]. Increased parasympathetic activity tends to take place during periods of relaxation and rest and even with structured meditation techniques. This results in overall lower heart rate variability. This typically causes an **overall reduction in autonomic nervous system outflow**. Increased parasympathetic activity also causes large peaks to take place in the HF band of the power spectrum. Coherence is associated with increased parasympathetic activity and is part of the relaxation response. When the body is at rest the parasympathetic and sympathetic nerves are tonically active, with vagal effects being predominant. Also meditation and relaxation do not always cause significant increases in coherence. This is because fundamental differences exist between coherence and the physiological state of relaxation. Also associated psychological states during these times can be markedly different.

A long term study that took place when geomagnetic activity was quiet found strong positive relationships between cosmic rays and HRV variables. This suggests HRV responds to increases in cosmic rays [4].

Summary

Coherence is associated with increased parasympathetic activity and is part of the relaxation response. Increased parasympathetic activity tends to take place during periods of relaxation and rest and even with structured meditation techniques.

References. Chapter 25.

1. McCraty, R., A. Deyhle, and D. Childre, The global coherence initiative: creating a coherent planetary standing wave. Glob Adv Health Med, 2012. 1(1): p. 64-77.
2. Dekker, J.M., et al., Heart rate variability from short electrocardiographic recordings predicts mortality from all causes in middle-aged and elderly men. The Zutphen Study. American Journal of Epidemiology, 1997. 145(10): p. 899-908.
3. Tsuji, H., et al., Reduced heart rate variability and mortality risk in an elderly cohort. The Framingham Heart Study. Circulation, 1994. 90(2): p. 878-883.
4. Siegman, A.W., et al., Dimensions of anger and CHD in men and women: self-ratings versus spouse ratings. J Behav Med, 1998. 21(4): p. 315-36.

Chapter 26. Final Conclusions and Summary

Persinger suggests that the space occupied by geomagnetic fields stores information that relates to brain activity and that this information is able to be accessed by the human brain [1]. Hence we have a basic theory of how consciousness operates.

Because earth's magnetic fields are carriers of biological information and because humans have heart and brain frequencies which constantly overlap with earth's magnetic field, not only is the human body a receiver of biological information, but these frequencies couple information to earth's magnetic fields, feeding a constant stream of information into the global field environment.

Experimental solid and conclusive evidence now exists showing human bio-emotional energy has a subtle, yet significant (scientifically measurable) nonlocal effect on organic matter, people and events [2]. Hence this bio-electromagnetic field, such as the field radiated outwards by the human heart and brain affects others and the "global information field environment" that we all reside in. To put this simply, we affect the global information field through intent and / or strong emotional feelings.

Recent research by the HeartMath Institute has confirmed the hypothesis that individuals in a state of coherence, radiate coherent electromagnetic signals out into the environment via their heart and while doing so, become more sensitive to detecting information contained within the fields radiated by others [3]. This detection may not be limited to the energy fields radiated by others, but also by energies that exist in other dimensions, including time and space. Further research is needed to confirm this. Growing evidence is starting to show that an energetic field forms among participants in groups throughout which communication among all the group members takes place simultaneously. To put this simply, an actual "group field" forms, connecting all its members [4]. In a noisy room filled with many conversations, for instance we have the ability to tune out the noise and focus on a single conversation of interest. Hence our mind has an ability to determine what gets processed at higher levels.

Further evidence supporting the hypothesis that magnetic fields act as carriers of biologically important information comes from a study conducted by Montagnier et al [5]. Montagnier found epigenetic information related to DNA was able to be detected in the form of electromagnetic signals that were channeled into a highly diluted solution. He discovered that this information was able to be transferred to and imprinted in pure water which had never been exposed to DNA. Also this information instructed the re-creation of DNA when **extremely low electromagnetic frequency fields** of 7.8 hertz were played in the presence of the basic constituents of DNA. Montagnier also found that the **presence of a magnetic field was necessary** for the information transfer to take place. Montagnier also stated that the very low electromagnetic frequency field which transfers DNA information may originate from natural sources such as the Schuman resonances (SR) (7.83 hertz).

This is yet another groundbreaking discovery because as mentioned earlier, a broken or injured plant leaf must have a source or method to receive the information to repair itself. If indeed DNA is acting as a receiver of information in the presence of 7.8hz, it may mean that the Schuman resonance, when at favourable levels, acts as a carrier or transmits information via various energy fields which the DNA of biological organisms pick up. These energy fields contain within them the information necessary for the plant to repair its leaf.

Final Hypothesis

As overall global coherence manifests itself during favourable solar weather, it affects consciousness perhaps by enhancing the rate / speed at which information is processed. Overall global coherence results in stronger galactic coherence within our   solar   system.

To use an illustration of the varying rate of the flow of time, if a person leaves earth in a rocket ship travelling at the speed of light and returns to earth a few months / years later, the clock on their ship will have run slower because of the speed at which the ship was travelling (the ship was in a moving bubble). Hence the ship travelled forward in time. Because information may exhibit non-local properties, during times of sustained coherence, conscious particles may travel in a non-local fashion further and faster, creating a stronger link to future information, When a person's mind is remote viewing the future, and their consciousness returns to the present, these particles of future information flood their awareness / consciousness giving them future information. Further research is necessary to prove this theory.

The Emerging Global Mind

Large numbers of people focused on generating heart-entered states of love, care and compassion generate a more intense coherent field that benefits others and offsets the current

planetary wide incoherence and discord.

Research studies now confirm that interactions between human emotions and a global field occur when large numbers of people perform organized global peace meditations or experience similar emotional responses to certain events [6 to 8]. Examples include where peaceful meditators were able to create a 25% drop in crime via group meditations [9]. The opposite example is 911 where earth's magnometer showed extreme activity, not only before the planes struck the twin towers in New York, but also during the event [10]. Hence the global field not only exhibited pre-stimuli effects, but was influenced in both cases in response to periods of strong focused emotional intent. The research study performed during 1993 in Washington, DC [11] showed that even a relatively small group of a few thousand can influence the larger whole (the D.C. Population).

A similar experiment was conducted during the peak of the Lebanon-Israel war during the 1980s by Drs. John Davies and Charles Alexander of Harvard University. They organized groups of professional meditators in Yugoslavia, Jerusalem and the United States and instructed them to meditate and focus their attention in specific regions over a 27-month period. The study found that levels of violence in Lebanon decreased 40% to 80% every time the meditating group meditated on quenching the violence. The largest reductions took place when there were more meditators. The study found that the average number of people killed per day fell from 12 to three. This is a decrease of more than 70%. Also war-related injuries fell by 68% and the intensity level of conflict level (which is a measure of violence caused by war) decreased by 48% [12, 13]. Hence, the intentional emotional violence put into earth's fields during a war, can be neutralized by organized groups of professional meditators focusing on emotions of love, peace and harmony.

When a person experiences feelings of compassion, they shift into a more coherent physiological state [14]. It is during this state they are radiating stronger magnetic waves of energy which radiate outwards into their environment [15]. Emotional states that generate compassion bring people together, making them coherent. Hence this also means that exhibiting compassion the days ARV session takes place may enhance the over collective coherence in the region one performs the ARV session in.

One interesting point regarding the D.C. Meditator Experiment, which took place between June 7 and July 30, in 1993, took place during a **long-term solar cycle condition green** (www.ez3dbiz.com/in_depth.html).

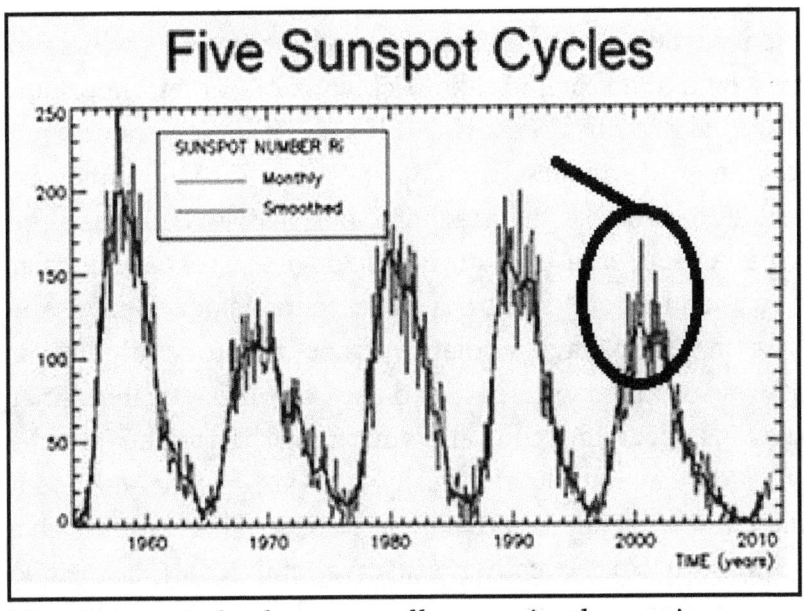

This is a period where overall peace in the environment is

more common, especially on a worldwide level due to declining solar activity and overall lower solar wind speeds. Hence the meditators mediating on peace + the condition green created a co-synergy between the earth and meditators enhancing the effects of peace and love in the environment.

RNG's.

Because major global events elicit strong emotion from the world's population, we can measure this impact of emotion using Random Number Generators (RNGs) [16], thus giving us a pictorial view of the global collective unconscious in real time. In one study, a multiple independent analyses of RNG's during the 911 terrorist attacks in New York on the morning of Sept. 11, 2001 showed large, significant shifts occurring in the output of the global network of RNGs [17].

In the future we may see the emergence of a device that neutralizes negative emotions by creating coherence. This device would activate fields of coherent energy, which in turn would allow one to take back control of their emotions. This can already be done using self-regulation techniques, but once the entire process is completely understood, it may be possible to assemble such technologies. This would make the device especially useful during stronger, unexpected solar activity and those suffering from depression. Hence the device may also have a solar weather monitor built into it that warns a person that stronger or unfavourable solar activity is occurring, thus allowing him / her to activate the device or have it automatically activate in the event of unfavourable solar weather conditions. This would give warnings to solar weather sensitive individuals or people suffering from ill health where they can then take the necessary countermeasures such as a change in diet or self-regulation techniques.

IMF Polarity

A research study looked at GCP data between the years 1998 and 2008 and matched satellite-based interplanetary magnetic field (IMF) polarity with GCP-world events. These events included celebrations, meditations, natural catastrophes or violence. The study discovered that RNG deviations depend on a positive IMF polarity which coincides with emotionally significant conditions and/or entropy changes [18]. This suggests that a relationship exists between interplanetary magnetic field (IMF) polarity and emotions, perhaps intensifying them in some manner. Further research is necessary to explore this further.

What is Positive IMF Polarity?

The interplanetary magnetic field (IMF) is part of our Sun's magnetic field which is carried into interplanetary space via the sun's solar wind. As it flows into space, the interplanetary

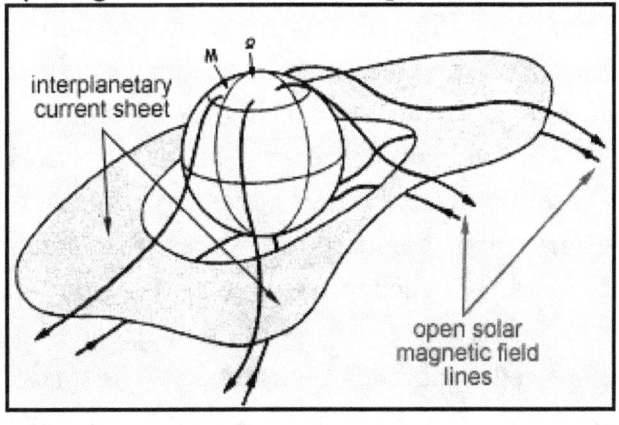

magnetic field lines become "frozen into" the plasma of the sun's solar wind. Due to the Sun's rotation, the IMF travels outward in a spiral pattern, just like the solar wind.

The IMF comes from regions on the Sun where the magnetic field is open. This means field lines that emerge from one region do not return to a conjugate region but instead head out non-stop into space. The direction, or polarity of this field in the Sun's northern hemisphere is opposite of the field in the Sun's southern hemisphere. This means that the polarities reverse each solar cycle. This is an interesting finding because as we showed earlier in this book during Etzold's second series of tests the influence of PSI appeared to have changed direction (reversed). Etzold (2005) discovered that solar activity (Solar Minimum and Maximum) was responsible for this reversal effect.

Future Plants of the Global Coherence Project

One of the goals of the HeartMath Institute is to test the hypothesis that large groups of heart-coherent people holding a shared intention are able to encode physiologically patterned and relevant information in earth's energetic and geomagnetic fields. Hence numerous studies already show the effects of meditation, prayer and groups sending out their intentions by using varying experimental contexts [19 to 21]. Also individually generated coherent waves are much more likely to couple themselves to a larger collective field environment compared to waves from states of incoherence.

The GCI theory states that as a large enough number of individuals increase their personal coherence that it causes increased social coherence (teams, families and organizations). As increasing numbers of social units (schools, families, communities, etc.) become more coherent it leads to increased overall global coherence. All this is enabled and advanced via self-reinforcing feedback loops that exist between humanity and the global field environment.

It takes shifts in consciousness to achieve new levels of cooperation and collaboration. Unfortunately many of these shifts take place during events involving major negative trauma. Research now shows that these shifts in consciousness can occur via coherence and self-regulation techniques, especially in groups. This opens new ways of solving problems and utilizing intuitive discernment. These methodologies are required if we are to honestly address our environmental, social and economic problems. The only thing lacking is the motivation. If enough social groups and individuals increase their coherence and are utilizing it to intentionally create a clear standing reference wave that is coherent to the overall global field, it lifts global consciousness as a whole in the process. This can only be achieved when increasing numbers of people move towards mastering self-mastery (know thyself). This in turn helps promote intuitive discernment and collaboration and cooperation for addressing today's problems.

Those who understand power, do not exercise power over others, instead they have the power to control their emotions and their ego.

References. Chapter 26.

1. Persinger, M., On the possible representation of the electromagnetic equivalents of all human memory within the earth's magnetic field: Implications of theoretical biology. Theoretical Biology Insights, 2008. 1: p. 3-11.

2. McCraty, R., A. Deyhle, and D. Childre, The global coherence initiative: creating a coherent planetary standing wave. Glob Adv Health Med, 2012. 1(1): p. 64-77.

3. McCraty, R., The energetic heart: Bioelectromagnetic communication within and between people, in Bioelectromagnetic Medicine, P.J. Rosch and M.S. Markov, Editors. 2004, Marcel Dekker: New York. p. 541-562.

4. McCraty, R., Childre, D, Coherence: Bridging Personal, Social and Global Health. Alternative Therapies in Health and Medicine, 2010. 16(4): p. 10-24.

5. Montagnier, L., et al., Transduction of DNA information through water and electromagnetic waves. arXiv preprint arXiv:1501.01620, 2014.

6. Davies, J.L., Alleviating political violence through enhancing coherence in collective consciousness: Impact assessment analysis of the Lebanon war. Dissertation Abstracts International, 1988. 49(8): p. 2381A.

7. Hagelin, J., The Power of the Collective. Shift: At the Frontier of Consciousness, 2007. 15: p. 16-20.

8. Hagelin, J.S., Orme-Johnson, D. W., Rainforth, M., Cavanaugh, K., & Alexander, C. N. , Results of the National Demonstration Project to Reduce Violent Crime and Improve Governmental Effectiveness in Washington, D.C. Social Indicators Research, 1999. 47: p. 153-201.

9. Hagelin, J.S., Orme-Johnson, D. W., Rainforth, M.,

Cavanaugh, K., & Alexander, C. N. , Results of the National Demonstration Project to Reduce Violent Crime and Improve Governmental Effectiveness in Washington, D.C. Social Indicators Research, 1999. 47: p. 153-201.

10.   REF magnometer disturbance on sept 11, 2001. [383-385] .

11. Hagelin, J.S., Orme-Johnson, D. W., Rainforth, M., Cavanaugh, K., & Alexander, C. N., Results of the National Demonstration Project to Reduce Violent Crime and Improve Governmental Effectiveness in Washington, D.C. Social Indicators Research, 1999. 47: p. 153-201.

12. Davies, J.L., Alleviating political violence through enhancing coherence in collective consciousness: Impact assessment analysis of the Lebanon war. Dissertation Abstracts International, 1988. 49(8): p. 2381A.

13. Orme-Johnson, D.W., et al., International Peace Project in the Middle East The Effects Of The Maharishi Technology Of The Unified Field. The Journal of Conflict Resolution, 1988. 32(4): p. 776-812.

14. McCraty, R., Atkinson, M., Tomasino, D., & Bradley, R. T, The coherent heart: Heart-brain interactions, psychophysiological coherence, and the emergence of system-wide order. Integral Review, 2009. 5(2): p. 10-115.

15. McCraty, R., The energetic heart: Bioelectromagnetic communication within and between people, in Bioelectromagnetic Medicine, P.J. Rosch and M.S. Markov, Editors. 2004, Marcel Dekker: New York. p. 541-562.

16. Bancel, P., Nelson, R., The GCP Event Experiment: Design, Analytical Methods, Results. Journal of Scientific Exploration, 2008. 22(3): p. 309-333.

17. Nelson, R., Effects of Globally Shared Attention and

Emotion. Journal of Cosmology, 2011.

18. Wendt, H.W., Mass emotions apparently affect nominally random quantum processes: interplanetary magnetic field polarity found critical, but how is causal path?, 2002, Halberg Chronobiology Center, University of Minnesota: St. Paul.

19. Ameling, A., Prayer: an ancient healing practice becomes new again. Holist Nurs Pract, 2000. 14(3): p. 40-8.

20. Gillum, F. and D.M. Griffith, Prayer and spiritual practices for health reasons among American adults: the role of race and ethnicity. J Relig Health. 49(3): p. 283-95.

21. Schwartz, S.A. and L. Dossey, Nonlocality, intention, and observer effects in healing studies: laying a foundation for the future. Explore (NY). 6(5): p. 295-307.[390-392]

Chapter 27 . A Brief outline of the RetroPschokinesis Project (RetroPK)

The published paper titled: Solar Periodic Full Moon Effect in the Fourmilab Retropsychokensis Project, which was published by Eckhard Etzold is a metadata analysis of psychokensis experiments. His paper includes what moon phases and solar activity were present when psychokensis experiments were taking place.

The data from a random number generator based on radioactive decay rates was made available to viewers using the world wide web. Visitors were asked to shift a series of random number generated data using their minds. Methods of influence suggested were wishing, prayer, wanting or intent. The results were than stored on a computer server. Many of these retropsychokensis experiments were conducted online from participants around the globe for over a number of years. While this study does not directly relate to associative remote viewing data, there exist similarities in the data that corresponded to ARV sessions. Especially the research conducted in the 1960's by Andrija Puharich who stated that telepathy experiments were strongest during times that the moon was full and new (Puharich, 1973). I sincerely believe that telepathy and associative remote viewing share a very close relationship, which makes research on telepathy a valuable asset when researching associative remote viewing. Also associative remote viewing involves drawing a picture of a future event, which involves will and intention, both of which are forms of psychokinesis. Let's examine the main highlights of Eckhard Etzold's paper in detail and look for correlations in our data and use it to confirm some of our hypothesis. Then we will then have a guide-map with which to look for cycles and patterns to enhance our associative remote viewing sessions.

The Full Moon

Peak effects of psychokinesis have been found to occur during full moons as stated by Radin and Reban (1998). This peak effect is believed to occur due to the "charge" effect that occurs on the moon's dark side as it passes through earth's magnetosphere during full moons. This charge is greatest 1 day before and after a full moon. The charge is much less on the moon's bright side due to the electromagnetic radiation / light of the sun behaving as a type of shield.

The Full Moon and Casino Winnings

During 1998 Radin and Rebman (1998 p. 193) found that the payout rate in casinos (from slots, craps, kenno, roulette and blackjack) showed a peak within 1 day of the full moon. After looking at this data a number of retropsychokensis experiments were conducted to verify and duplicate this effect (Solar-periodic full moon effect in the fourmilab retropsychokensis project. Eckard Etzold). After a series of tests over a number of years it was found that a cycle existed that was dependent upon solar activity and lunar phase. To put it simply, when solar activity was decreasing, the effects of the full moon during retropsychokensis experiments became maximized. When solar activity begin increasing, retropsychokensis experiments conducted during the full moon were not as successful. Also studies by Andrija Puharich (1973. pgs 281 to 289) found that telepathy experiments worked best during full moons and decreased at the quarter moons than increased again during new moons. Gaming winnings occurred most often around full moons (Radin 1987. The Conscious Universe. Chapter 11).

As the interval of the moon fades, so to does its effects. Experiments by Rebman and Radin (1998. p. 208) found that the interval 2 to 3 days around a full moon found 3 payout rates of casino games was present.

Our research at the Solar Institute early on found that a relationship does in-deed exist between full moons and enhanced accuracy during associative remote viewing sessions, but we overlooked the existence of **stronger solar activity during full moons**, which reduced ARV accuracy. This effect is very similar to how cosmic rays enter earth's atmosphere. When solar activity is low, more cosmic rays enter earth because solar activity provides a type of shield that repels cosmic rays. This is why it rains more when solar activity is low as more cosmic rays create more clouds. When solar activity is strong, cosmic rays bounce off the abundance of solar electrons. This same effect may be taking place when the moon is full and solar activity is stronger. It may also explain why our ARV sessions failed when a solar flare had occurred within 72 hours of an ARV session and solar wind speeds were above 400, as solar flares provide more electrons which may be shielding the beneficial effects of the moon, as the moon phase has been shown to be a deciding factor in the success of RetroPK experiments. While this may seem like a repeat, it is a critical finding that is easily overlooked in all the data and is well worth exploring further.

Dorman and Shatashvili (1961) observed during full moons that the secondary cosmic ray flux (SCR) increased, however during new moons, a decrease took place.

Thunderstorms and the Full Moon

The Schuman resonance is caused by lightening, which usually occurs during thunderstorms (Application of the Schumann resonance spectral decomposition in characterizing the main African thunderstorm center. Michal Dyrda et al. Oct 2014) and during full moons thunderstorms are more frequent, especially the following two days after the full moon and that the increase may be due to earth's magnetotail (Relationship between Thunderstorm Frequency and Lunar Phase and Declination. 20 September 1970).

Solar Cycles and PK

During solar maximum our sun will produce on average of three CME's each day. During solar minimum about one CME takes place on average every five days (Coronal Mass Ejections. Fox, Nicky. NASA/International Solar-Terrestrial Physics. 6 April 2011). This effect was noticed in RetroPk experiments conducted during the Summer of 2000. After numerous successful RetroPK trials that took place during the full moon from 1997 to 1999 during the year 2000, the effects appeared to reverse themselves (the psi missing effect). The year 2000 happened to be during a peak / accumulation of solar activity, when solar activity began entering its declining phase (change in polarity so to speak). This was where solar activity went from peaking to declining.

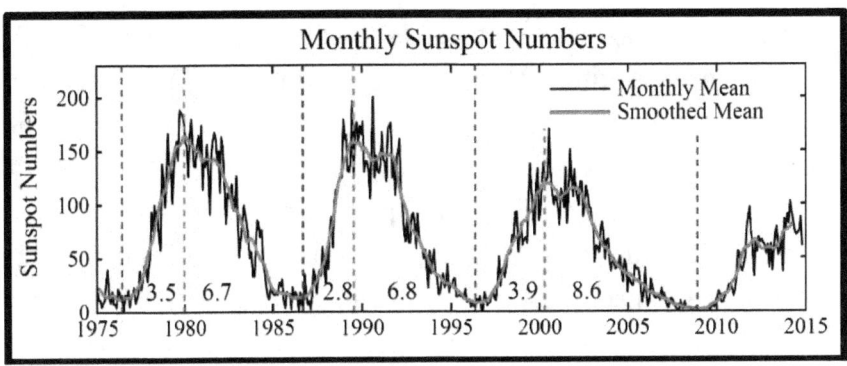

Hence we see the beginning of a cycle of ARV related phonenomena. To put it simply, successful ARV sessions are more common when solar activity is in decline, especially when it is reaching its low. Hence when planning ARV sessions it would be best to know what solar cycle is present in order maximize results.

The existence of these two types of reversing cycles / opposing polarities which are wholly influenced by solar activity may also provide an explanation as to why specific anti-aging visualization exercises work well at some times and not at others. It may also explain why people born during solar maximum live 5 years shorter on average (sorry millennials) compared to people who are born during solar minimum (Solar Activity at Birth Predicted Infant Survival and Women's Fertility in Historical Norway. Gine Roll Skjærvø et al. Jan 2015).

An interesting observation I have made over the years is when the sun's 2MEV levels are stronger, especially when peaking for a number of days, the ability to heal the body is significantly enhanced. For example I wrote an article in 2018 titled: Exploring the Emma McKinley Healing Miracle. In this article I showed her healing took place when the solar wind speed was at a favourable speed and the sun's 2MEV levels were peaking. Above average peaks in 2MEV activity usually take place just after higher solar activity (the condition green period).

So in summary the same power of the mind used to remote view future events, may be a promising candidate for healing and anti-aging as the power of the mind is stronger during these specific time periods / cycles. Also the HeartMath rejuvenevity exercise seems to work best at a half / first quarter moon (or recovery from exercise). This effect may be enhanced when solar wind speeds are at around 330 and solar activity is quiet.

Photosynthesis and Quantum Mechanics

There exist parallel earths and universes all about us. Each one contains a slightly different version of you. For example in one universe you might be president of the United States, in another universe a physics teacher and in another a garbage collector. The activity of photosynthesis, which exhibits quantum effects [3] may be able to teach us more about parallel universes then we think. The rule of photosynthesis is that out of multiple pathways that electrons choose to travel to, they will always follow the path that has the least resistance (superconductor behaviour). Hence multiple worlds are accessed, sensed via gateways / pathways of least resistance. The more favourable the solar weather, the less resistance.

ULF Waves and PK

Further research by Etzold (2005) found that during full moons, four main variables would come into play which affected RetroPK performance. These were sunspots, ecliptic angle, F10.7cm and geomagnetic activity. These variables showed no major effect during the other moon phases. Etzold (2005) goes on to state that RetroPK effects may be caused by the moon entering earth's magnetosphere while full, perhaps by emitting ULF (ultra-low frequencies) waves.

Studies conducted by Dimitrova and Khabarova found that ULF waves between 2–10 MHZ exhibited strong correlations with a rise in blood pressure (0.6) compared to geomagnetic measures (0.3) (Some Proves of Integrated Influence of Geomagnetic Activity and Weather Changes on Human Health. Khabarova O.V. and Dimitrova S. Sept 2008).

Zenchenko et al. found that in two-thirds of their studies that synchronization is taking between human heart rhythms and ultra-low frequencies (0.5 to 3.0 MHZ) which exist in earth's geomagnetic field (Synchronization of human heart-rate indicators and geomagnetic field variations in the frequency range of 0.5–3.0 MHZ. T. A. Zenchenko. Dec 2014). What is interesting is the frequency of some cosmic ray showers occur at the frequency of 2MHZ (Evidence of Radio Pulses at 2 MHz from Cosmic Ray Air Showers. T. J. Stubbs. April 1971) and the frequency of 2.5 mHz is used to measure Ultralow frequency (ULF) waves in earth's magnetosphere and its relationship to the sun's solar wind (ULF Wave Activity in the Magnetosphere: Resolving Solar Wind Interdependencies to Identify Driving Mechanisms. S. N. Bentley et al .March 2018).

Further Reading

Synchronization of Human Autonomic Nervous System Rhythms with Geomagnetic Activity in Human Subjects.

Rollin McCraty. July 2017.

Summary
Frequencies between 0.5 and 3.0 MHZ appear to take place in earth's upper atmosphere. During favorable solar weather conditions, these low frequencies that are in the ULF range are most beneficial, enhancing the results of ARV sessions.

Speaking from personal experience, I have found that while listening to a frequency recording of the sun (3.0mHz) that is has a powerful rejuvenative / refreshing effect upon the body, especially physical strength. You can listen to the recording made by Stanford University Solar Studies Center on Youtube by watching the video titled: **Sound of our Sun. Filtered Version.** The frequency of this is approximately 3.0 MHZ.

Cosmic Rays and the Sun's 10.7cm Radio Flux
As discussed earlier in this book, we saw that an increase in the sun's 10.7cm radio flux was associated with an increase in feelings of well-being and happiness (www.HeartMath.org. Interconnectedness. October 6, 2010. GCI Commentaries. Science Study. www.heartmath.org/gci-commentaries/interconnectedness.).

Shown on pages 246 and 250 of the report titled: **The Cosmic Ray and the 10.7 Cm Flux Variations During Solar Cycles 19-23** that was published by J. E. Mendoza-Torres. et al. in 2014, showed that the sun's 10.7cm solar radio flux shows a cycle occurring with cosmic rays. When the sun's **10.7cm solar radio flux increases, so do cosmic rays.** The following image is from the study, I have drawn in lines to show the rise in both variables. The sun's 10.7 cm radio flux is shown at the top with the cosmic ray count at the bottom.

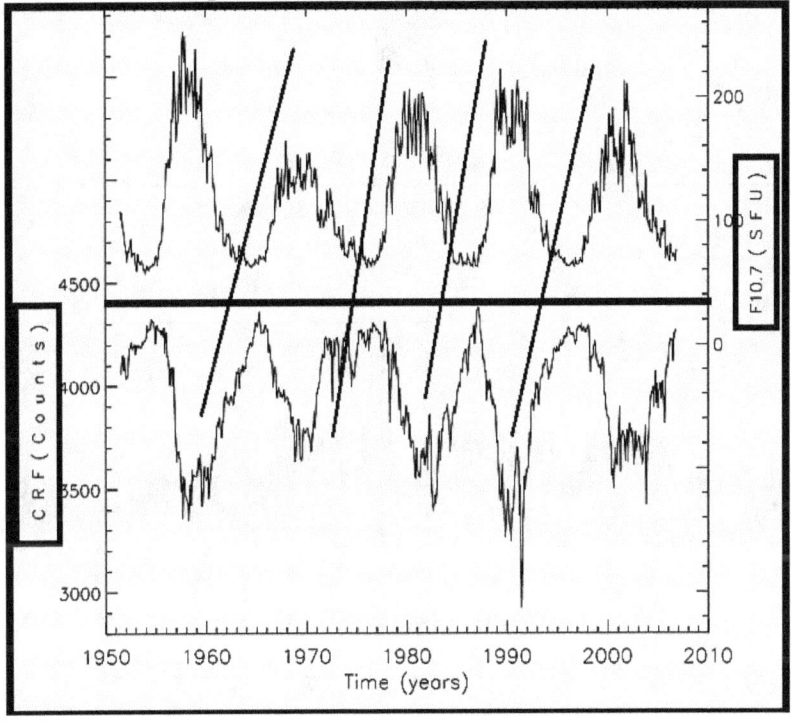

Even though stronger solar activity repels cosmic rays, because more sunspots act as a type of shield against cosmic rays (The Solar Cycle. David H. Hathaway. Sept 2015), one would think that an increase in the sun's 10.7cm radio flux would repel cosmic rays because enhanced solar activity is usually associated with an increase in the sun's 10.7cm solar radio flux, but this is not always the case and may have to do with solar wind speed. The next image shows the sun's 10.7cm solar radio flux in proportion to long-term solar cycles. The following image shows slight rises taking place in the sun's 10.7cm solar radio flux near sunspot maximum.

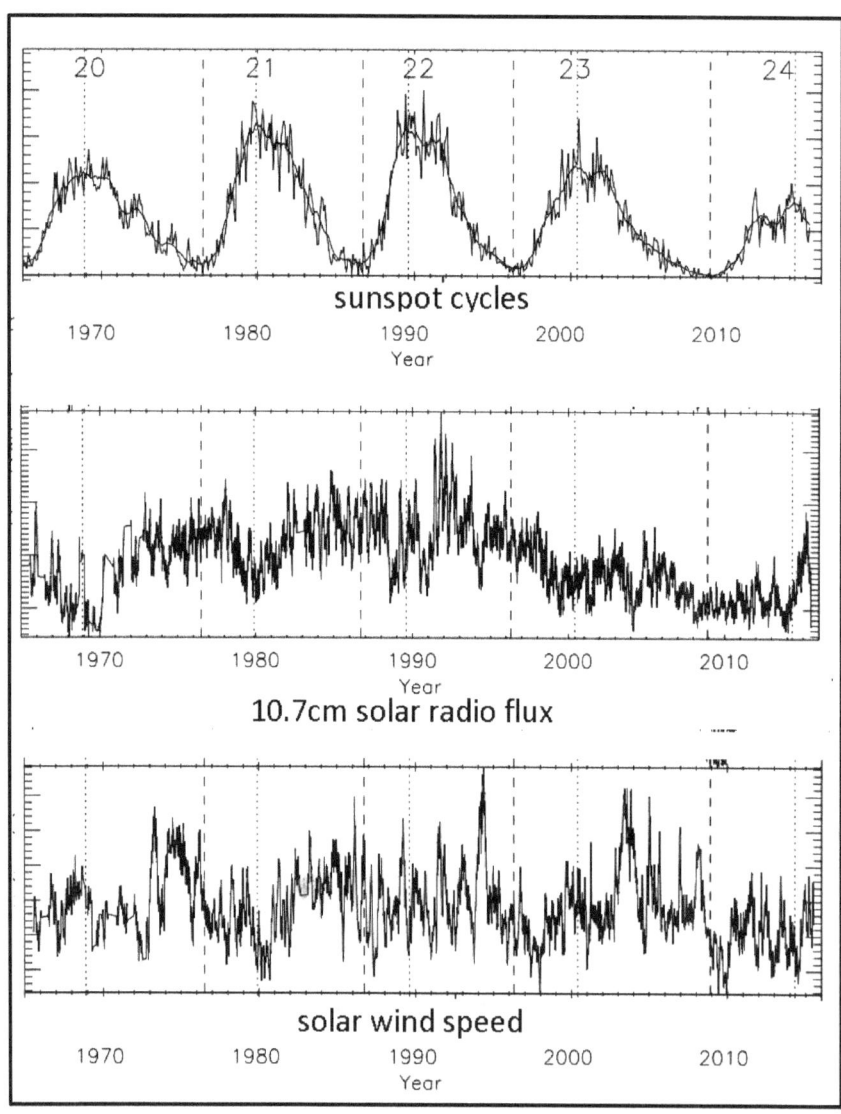

sunspot cycles

10.7cm solar radio flux

solar wind speed

Reference

The Solar Wind During Current and Past Solar Minima and Maxima. J.L. Zerbo and J.D. Richardson. Dec 2015.

The Sun's 10.7cm Solar Radio Flux and Radar

The 10.7cm solar radio flux is used to calibrate antennas due to its sensitivity to the sun's chromosphere (The 10.7 cm solar radio flux (F10.7) K. F. Tapping. March 2013). The 10.7cm solar radio flux can also be picked up on radar. During WW2, radar operators observed what they called "sun strobes" (also called increased receiver noise levels) that would appear on their planned position indicator displays whenever their antennas scanned across the azimuth of earth's horizon. This phenomenon would only take place when the sun was directly on the horizon and setting or rising (The 10.7 cm solar radio flux (F10.7) K. F. Tapping. March 2013).

In Ottawa Canada, Arthur Covington and colleagues spent the remaining war years using components from surplus radar systems to design Canada's first radio telescope which used an operating wavelength of 10.7 cm (**2.8 GHz**). The 10.7cm radio flux follows a 10–13 year cycle with the solar activity cycle. Large variations take place around solar maximum due to the decay and appearance of active regions (The 10.7 cm solar radio flux (F10.7) K. F. Tapping. March 2013).

Theory

Perhaps an increase in cosmic rays during an increase in the sun's 10.7cm solar radio flux creates a type of reduced interference of some kind as radar has shown that it increased the noise. I show in our third edition of the associative remote viewing series titled: Improve Your Remote Viewing Accuracy Techniques Using Quantum Microtubules, that a specific type of noise actually enhances remote viewing. This effect may also have something to do with what's known as gyro-synchrotron emission which is the electromagnetic radiation that is emitted by charged particles that are moving at close to the speed of light in a magnetic field. Further research is necessary to explore this further.

Summary

The increase in PSI or psychic ability may be partly attributed to enhanced cosmic rays with the strongest peaks occurring at approximately the same time of day. These cosmic rays also show a trend that rise and fall with the sun's 10.7cm solar radio flux. Because cosmic ray activity is forecast to reach record highs in the coming years, which in turn leads to enhanced intuitive abilities, we may also expect to see more secrets revealed.

The success of PK is governed by moon phase and solar activity. At the right times a beneficial synergistic reaction occurs (full moons and quiet solar activity). This results in a layover of frequencies, thus creating stronger fields of energy. This is not unlike the twin participant experiment mentioned earlier in this book where the study used 1 person looking at a monitor and two participants watching pictures on a monitor while they faced one another. The study found that the co-participant pair in the experiment showed a much stronger pre-stimulus response, compared to just a single person.

It is interesting to note that our ARV device makes use of a small DC motor that is spun while the piezoelectricity is activated which acts as a mini-generator that lights a LED light. The magnetic field generated by the small D.C. Motor may be generating an electromagnetic field similar in scope to the electromagnetic field generated by the heart. As the previous study showed, two people in coherence achieved better results than a single person alone, and a person in coherence is generating an electromagnetic field via their heart.

Summary
Dual electromagnetic fields in close proximity to one another generate overlapping frequencies which greatly enhance intuition / precognition.

Geomagnetic Activity During Full Moons Results in Less Lottery Winners

There is one factor that affects the moon's energetic fields as they impinge upon our earth and it is geomagnetic activity. Research by Radio (1997. p. 187) found that the lower payout rates from lotteries occurred during the time of the full moon when geomagnetic activity was at **ABOVE** average levels.

Summary
Above average geomagnetic activity, coupled with higher solar wind speeds, may be causing PSI interference when a full moon is present. This cause of this enhanced interference may be exhibiting / displaying a ladder / replicative type effect. If this is so, than the opposite would also be true, when a full moon is present and geomagnetic activity is low / quiet, with solar wind speeds in the range of 350, interference is greatly reduced. Hence, a major factor in successful ARV sessions is knowing the amount of geomagnetic activity, the solar wind speed and the lunar phase.
 Speaking from personal experience, the Solar Institute

has found the best geomagnetic levels to be between 4 and 11, as indicated by the Middle Latitude Fredericksburg K-indices. The address for real time KP activity can be found at: legacy-www.swpc.noaa.gov/ftpdir/indices/DGD.txt

The Piezoelectric Effect

Above average geomagnetic activity has also been linked to poltergeist activity and the hemolysis of red blood cells (Palmer, Baumann & Simmonds 2005, Braud & Dennis 1989, Gearhart and Persinger 1986). I first got the idea of introducing piezoelectricity into the ARV Amplification device after hearing about experiences by 'ghost busters' that hauntings happen to be more common in regions there is an abundance of piezoelectricity. If you were to do an Internet search for the term **hauntings + piezoelectricity**, you will find a number of researchers who have discovered this connection. One of the more popular cases is commonly known as The Stanley Effect: A Piezoelectric Nightmare.

Piezo1

Piezo1 is what's termed a mechano-sensitive ion channel protein which exists in humans that is encoded by the gene PIEZO1. Piezo1 occurs in the body's bladder, lungs and skin. Piezo1 also happens to be found in red blood cells. As we have shown in this book, the body's nervous system is connected with the bladder and in TCM the energy of the bladder peaks between **11 p.m. to 1 a.m with a seasonal peak occurring in December.**

What are Mechanosensitive ions?

Mechanosensitive is a method to describe Mechanosensory

transduction that governs the body's hearing and touch. A recent study identified Piezo1 and Piezo2 as a novel class of mechanosensitive channels.

Now that we have a more in-depth understanding of how geomagnetic activity and the moon interact with the human body, let's do a final review of the material in the next chapter.

Chapter 28. Cycles of Geomagnetic Activity and the Moon

Research has confirmed that geomagnetic activity on average always drops slightly before a full moon [1] [2] and increases just after a full moon. Because earth's geomagnetic activity declines just before full moons, this may create a synergistic type effect when RetroPK experiments are conducted (synergy between the enhanced electromagnetic fields that take place during a full moon [3] and lower geomagnetic activity). It has already been shown that the RNA expression in fish is affected strongest from the first quarter moon which than decreases before a full moon [4].

What is RNA Expression?

RNA Expression is a process through which information from genes are used in the synthesis of a gene function. In most cases they are proteins, but non-protein coding genes include RNA (tRNA) or small nuclear RNA (snRNA) genes.

Reversing Trends

The evidence that a long term moon / PSI cycle first existed in RetroPK experiments first took place during the summer of the year 2000 during the Solar Periodic Full Moon Effect in the Fourmilab Retropsychokensis Project [5]. As we covered earlier on in this book, when RetroPK experiments were conducted during full moons during the summer of the year 2000, the sessions were not as successful (Etzold. 2002a. p. 78). This same reversal trend / pattern was also seen in Radin's research (Radin, 1997. p.187). Hence the reversal effect could be associated with the long term 11 year solar cycle. If we look at a long term cycle of solar activity in the following image, we see that the year 2000 (solar cycle 24) was a phase in what's commonly called solar maximum. This is a period where solar activity has peaked. Hence the electromagnetic and increased UV activity from the sun may be shielding the positive "charge" effect caused by the full moon being in earth's magnetotail. This also results in reduced cosmic rays. As covered earlier, this is known as **ARV Cycle #1.**

Hence, it would make sense **ARV Cycle #2** (being the reverse of ARV Cycle #1) would take place at or near solar minimum, which just happens to be the period this book is being written (**Spring 2018**).

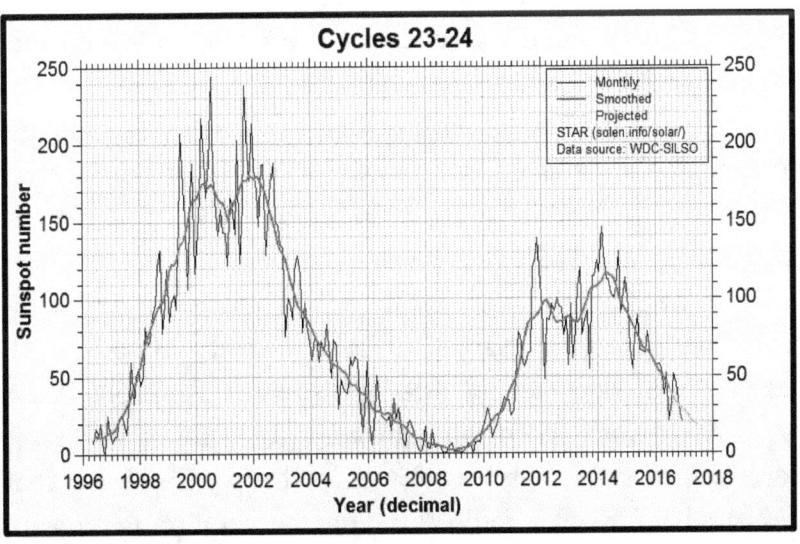

This would mean that the success of ARV sessions conducted during ARV Cycle #2 may be governed by phases other than full moons with solar wind speeds being more prominent.

Summary

The success of RetroPK is dependent upon lower geomagnetic activity taking place during full moons, and especially so when solar activity is quiet and cosmic rays are increasing. Stronger solar activity is usually accompanied by increased geomagnetic activity, which can be detrimental to RetroPK sessions. ARV effects are drastically reduced if a solar flare occurs during a full moon when solar wind speeds are above 370. We have found this to be true when we performed an ARV session during a full moon when there was a solar flare, which resulted in higher solar wind speeds. Hence, the ARV session   failed.

   An additional reason ARV sessions are less accurate during these periods is due to cosmic rays. Stronger solar activity creates a shielding type effect resulting in less cosmic rays. In addition, extra solar radiation emitted by the sun may be causing a shielding type from the moon's electromagnetic energy when the sun is more active.

Solar Activity and PSI

Research by Rebman and Radin (1998. pgs 212 and 216) found that specific variables of solar activity affected PSI. These types of variables included the sun's 10.7cm solar radio flux, solar wind speed and number of sunspots. It is interesting to note that we showed earlier in this book that the sun's 10.7cm solar radio flux affected people's feelings of well being. Also sunspots that take place along the sun's equator have more of an impact on earth compared to sunspots that occur at the sun's polar region and mid-latitude regions.

   Additionally, studies have shown that geomagnetic

activity affects human health and behaviour **at earth's higher latitudes** (Kalevi Mursula et al. Dec 2016. Seasonal solar wind speeds for the last 100 years). Hence it may be that the energy in orbiting bodies is released at higher latitudes or that people become more sensitive to the effects of solar weather at higher latitudes. Additional sunspot data can be found at RWC Belgium World Data Center for the Sunspot Index.

As shown throughout this book, from my own personal research, I have indeed found that the greatest impact to a successful ARV session is low solar wind speed combined with quiet geomagnetic activity. The beneficial effects were always greatest when the sun's 10.7cm radio flux (F10) levels are rising. An added boost is when the solar wind speed is at approximately 350 and there have been a series of mild solar flares having taken place the previous 4 days with the solar flares starting to fade out towards the date the ARV session is performed. Hence, maximum PSI effects occur as the moon moves through earth's magnetotail while full.

Now what is the prime leading mechanism for these effects? It is our theory at the Solar Institute that as the moon orbits the earth, it leaves a wake or trail behind it. This disturbance in space comes into contact with earth's magnetosphere creating a series of overlapping waves of various frequencies. During a full moon, the moon moves towards the middle of earth's magnetotail. The moon may be amplifying or sending the effects of these waves generated by the magnetotail closer into earth's ionosphere, which in turn is absorbed by the Schuman resonance. Further research will obviously explore this relationship in the future and the results should be rather interesting.

References. Chapter 28.

1. Dependence of the lunar modulation of geomagnetic activity on the celestial latitude of the Moon. Feb 1966.

2. Concerning a lunar modulation of geomagnetic activity. Aug 1964.

3. The impact of electromagnetic waves at the full moon on some physiological changes among the students of the Faculty of Sport Sciences at Mu'tah University. Dr. Baker Sulaiman Thuneibat. 2014.

4. Hypothalamic Expression and Moonlight-Independent Changes of Cry3 and Per4 Implicate Their Roles in Lunar Clock Oscillators of the Lunar-Responsive Goldlined Spinefoot. Riko Toda et al. Oct 2014.

5. Solar Periodic full moon effect in the fourmilab retropsychokensis project. Eckhard Etzold.

Chapter 29. Creating a Template for Remote Viewing the Financial Markets

The Basic Fundamentals of Initiating an Associative Remote Viewing Protocol for the FOREX and Dow Jones Markets

Creating the Framework

Before the remote viewing session is conducted, it is key to have a basic understanding of the market you are remote viewing. Just enough information is necessary to know what you are remote viewing, too much information can lead to confusion.

For example, one should know the hours the market is open and closed for trading and how much a given currency or stock moves during the week (trading activity). You should also associate yourself with the trading symbols about to be remote viewed. Once this is done, a template is created that is filled in during the ARV session. For example a template for the FOREX is shown in the following image. Because we remote view only as far as 4 days into the future, the template is designed to show where the financial currency or stock will be at the end date. Trying to fill in the actual trading activity over the course of the 4 days or so during an ARV session is not as accurate as aiming for where the trading activity will be on the actual end date. So it is best to focus on the exact date you are projecting to during an ARV session.

The following two images are templates that are filled in during an ARV session. The first is the FOREX template and the second is a Dow Jones ARV Template.

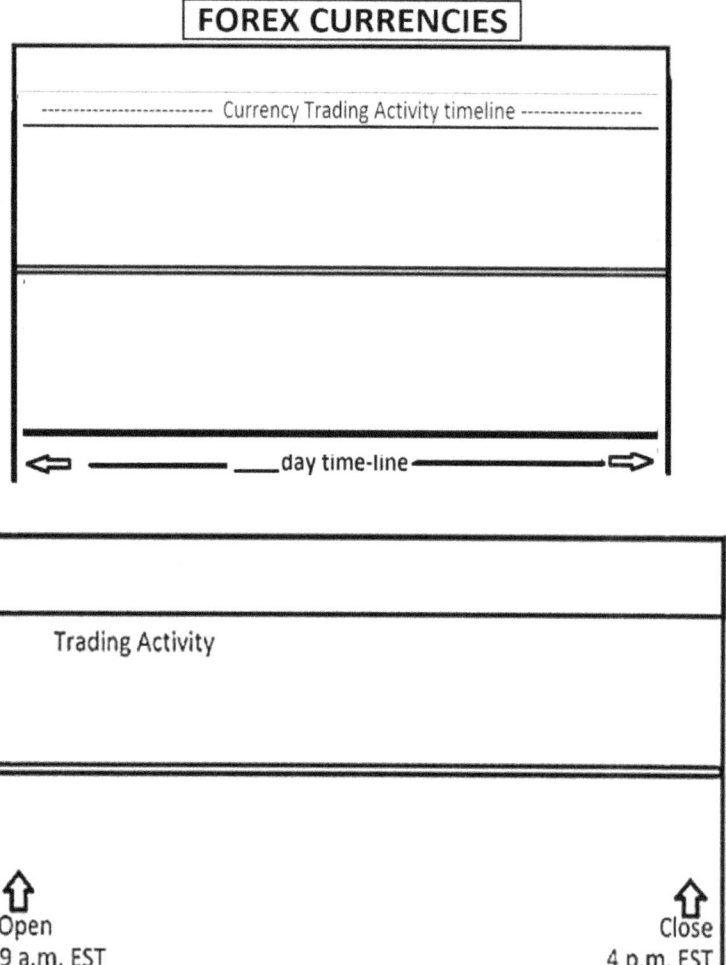

Dow Jones ARV Template

As stated earlier, it is also key to know when the public holidays related to the market take place so you don't perform an ARV session looking at a future date when the market is closed. I have done this before and I end up getting what looks like activity the next trading day.

An Introduction to Remote Viewing the FOREX.
Schumann Resonance Coherence Secrets.

An important point to consider is that a FOREX market is going to show a longer flow going from high to low or vice versa over the period of several days, compared to just a 24 hour period. We choose the FOREX for remote viewing because the future graph that is drawn during the ARV session is much easier to lookup online. For example when you input "dow jones chart" into google, the actual chart that occurs may show a downwards picture thinking the dow will close lower. So even if your dow jones remote viewing session showed a line closing lower, during the ARV session you may also get emotional impressions that it is going to close higher. This can result in conflicting information. In these cases it is just better to go with a market that uses a clear graph for the markets instead of the actual number.

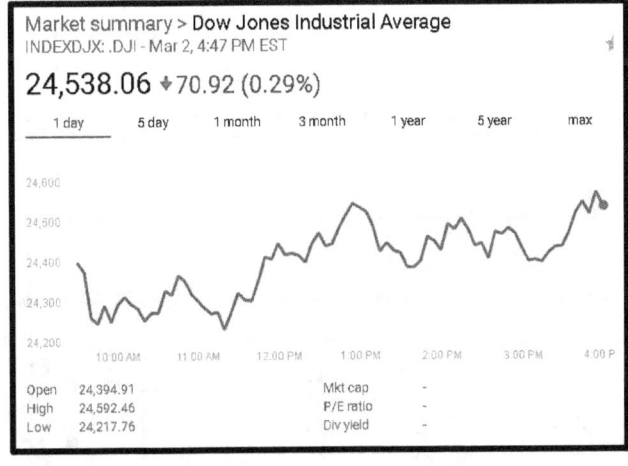

The image shown was displayed after entering the term "Dow Jones Chart" into the google search engine after one of our ARV sessions.

The unconscious mind responds much better to symbols and symbology rather than actual numbers. Because the best ARV sessions work well filling in a graph template showing the activity of the FOREX, it is why we choose to remote view the FOREX rather than the dow, as the displayed results of activity do not have the same effect as the Dow Jones.

Making Money on a Falling Market

You can make money when the currency falls instead of rises. This is known as an option. With an option you buy a contract stating that the currency is going to fall or decline within a certain range in the future and when the currency enters that range, you make money. Of course it's easier to make money when the currency goes from low to high and that is what we aim for during our ARV sessions.

Finding Favorable Solar Weather Conditions for an ARV Session

As we have covered extensively in this book, favorable solar weather conditions always take place just after a period of stronger solar activity. Solar minimums are the best time for ARV

#		Quarterly Daily So
#		
#		Sunspot
#	Radio SESC	Area
#	Flux Sunspot	10E-6
# Date	10.7cm Number	Hemis. R
#-------		-------
2017 01 01	73 0	0
2017 01 02	73 0	0
2017 01 03	73 11	0
2017 01 04	72 0	0
2017 01 05	73 0	0
2017 01 06	72 0	0
2017 01 07	72 0	0
2017 01 08	72 0	0
2017 01 09	71 0	0
2017 01 10	73 0	0
2017 01 11	75 0	0
2017 01 12	76 11	30
2017 01 13	75 24	70
2017 01 14	77 25	190
2017 01 15	78 23	180
2017 01 16	78 24	180

best time for ARV sessions. As shown earlier, this is ARV cycle #1.

Finding the "Sweet Spot"

The sweet spot is the period just after a geomagnetic storm during a full moon with favorable solar wind speeds. The previous image shows falling sunspots, which is usually accompanied by a decrease in geomagnetic activity and a less disturbed magnometer, as shown in the following image.

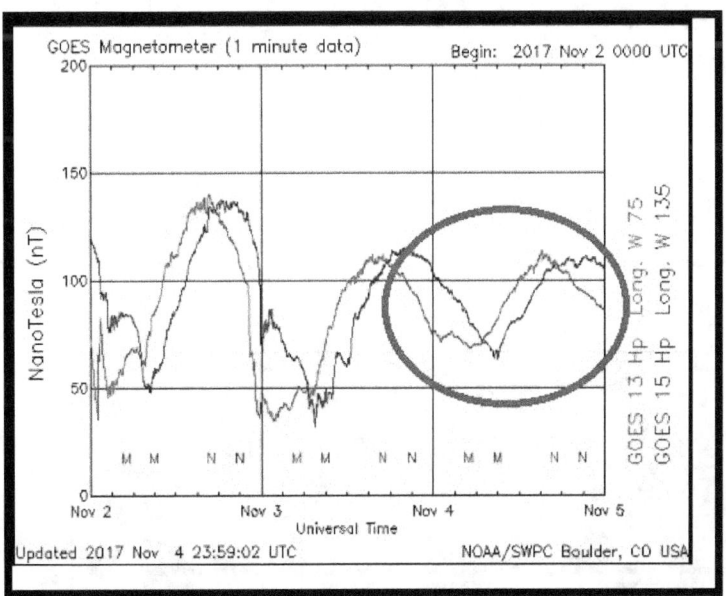

Narrowing Down the Sweet Spot

At the very beginning of stronger solar activity, the solar wind will "leap" followed by increased geomagnetic activity as shown in the following images.

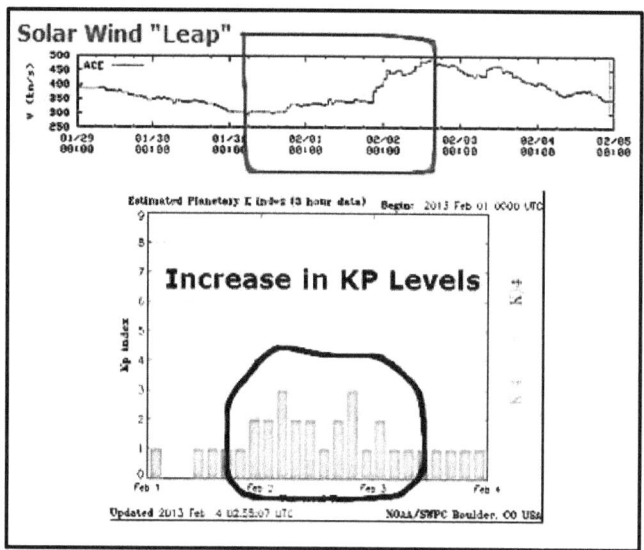

Next as solar activity begins to fade, Fredericksburg K-indices will begin to enter the sweet spot range of between 11 and 7 as shown below.

Date	A	Middle Latitude - Fredericksburg - K-indices							
2017 03 23	9	4	3	2	2	1	2	1	1
2017 03 24	7	2	3	2	0	2	2	2	1
2017 03 25	3	0	2	0	0	2	2	1	1
2017 03 26	3	1	0	0	1	1	1	1	2
2017 03 27	34	2	3	5	5	5	4	5	4
2017 03 28	-1	5	5	3	3	3	2	3	-1

Please note the optimal sweet spot is between 11 and 7. However during a full moon, the Fredericksburg K-indices will not always be in this range. As long as they are below 11, and solar wind speeds are favorable, the clarity of the ARV session will always be clear. Anything **above 11 and the success of the ARV session accuracy is greatly hindered**. The main key is to be sure the magnometer is not overly disturbed.

Solar Weather Forecasting Tools and Links

GEOMAGNETIC STORM FORECASTING

Canadian 27 day forecast
http://www.spaceweather.gc.ca/forecast-prevision/long/sflt-1-eng.php

NOAA Long Range Forecast
http://www.swpc.noaa.gov/products/27-day-outlook-107-cm-radio-flux-and-geomagnetic-indices

ADDITIONAL GEOMAGNETIC DATA

http://legacy-www.swpc.noaa.gov/ftpdir/latest/advisory-outlook.txt

KP 24 Hours Forecast
http://legacy-www.swpc.noaa.gov/ftpdir/indices/DGD.txt

http://legacy-www.swpc.noaa.gov/ftpdir/latest/daypre.txt

KP 3 Day Forecast
http://www.swpc.noaa.gov/products/3-day-forecast

SOLAR WIND

Wang Sheely Forecast
http://legacy-www.swpc.noaa.gov/ws/

An Introduction to Remote Viewing the FOREX.
Schumann Resonance Coherence Secrets.

Real Time Solar Wind Speed
http://legacy-www.swpc.noaa.gov/ftpdir/lists/ace/

Cal Tech Real Time Solar Wind Graph (SELECT 7 DAYS)
http://www.swpc.noaa.gov/products/ace-real-time-solar-wind

Real Time Solar Wind (black line is wind speed)
http://www.swpc.noaa.gov/products/real-time-solar-wind
Magnometer

Magnometer
Real Time Activity http://www.swpc.noaa.gov/products/goes-magnetometer

The Sun's 10.7 cm radio flux

Real Time Data
http://legacy-www.swpc.noaa.gov/ftpdir/indices/quar_DSD.txt

Lockheed Forecast F10 and AP Outlook
http://spawx.nwra.com/spawx/27do.html

Additional Data

0.8 MEV - 4 to 5 days before a sweet spot, there is usually a peak or increase in the 0.8 MEV particle photon flux.
http://legacy-www.swpc.noaa.gov/ftpdir/latest/DPD.txt

NOAA Solar Weather Forecast Data
http://legacy-www.swpc.noaa.gov/ftpdir/latest/forecast_discussion.txt

Cosmic Rays
https://cosmicrays.oulu.fi/

NOAA Warehoused Data
http://legacy-www.swpc.noaa.gov/ftpdir/warehouse/

http://legacy-www.swpc.noaa.gov/ftpmenu/indices.html

The final concluding chapters of this edition will re-cover Heart Rate Variability in further detail. I wanted to leave these chapters in this book because the science is so new. That said, I hope this book has helped look at how one can utilize these tools and new discoveries to create powerful and positive self-mastery in order to not only enhance intuition, but to have true and lasting health.

An Introduction to Remote Viewing the FOREX.
Schumann Resonance Coherence Secrets.

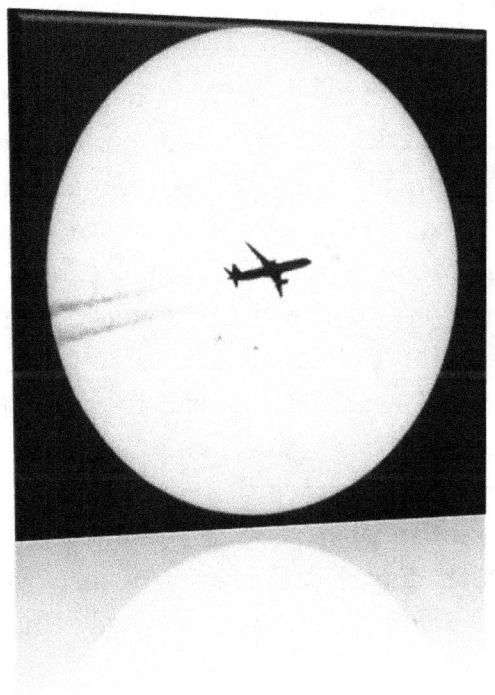

Chapter 30. Learning Self Regulation to Relieve Stress

As stated early on in this book, one of the major hurdles to a successful remote viewing session is for one's ability to control their emotions. Excessive stress may lead to severe emotional unbalances which can contribute to a type of psychic   interference.

Research studies have found that negative emotions and stress increase disease and can worsen the prognosis for individuals that already are suffering from an illness [1] [2]. In this chapter we are going to explore self-regulation techniques that one can use to gain control over their emotions, which are one of the major keys to self-mastery, and as a bonus, allow one to enjoy better long term health.

Today between 60% and 80% of visits to the doctor are stress related. What is alarming is that only 3% of patients receive any type of stress management help [3] [4]. Early heart research found that excessive physical effort or long term emotional influences which deprived the heart of rest, would make a person vulnerable to disorders and increase their vulnerability to disease [5].

Because nature works on the principle of dual polarity (the functioning of opposites / mirror effect), the opposite is also true. Positive emotions and self-regulation skills have been shown to significantly reduce premature mortality and prolong health [6 to 11]. This is a major finding, because it clearly shows that emotions impact our body's physical health and well being. By using self-regulation techniques, we now have the tools necessary to take back control of unwanted emotions that contribute to ill health [6 to 11]. Hence looking at this from a physiological perspective, our emotions are an important indicator as to how we deal with stress. Feelings of irritation, lack of control, anxiety, frustration and hopelessness are what we experience when we feel 'stressed out', especially if a situation may look as if it is beyond our control. These emotions could come from major life changes or just minor inconveniences. If small amounts of stress are allowed to accumulate, it will build up its own irritable resonance where even a small stressful situation may trigger an emotional outburst [12]. Hence, it makes sense positive emotions can be a strong predictor of good health. This applies even to those who may be without food or shelter. And long term negative emotions can be a strong predictor of bad health even if one's basic needs are met [13].

In a study involving elderly nuns, researchers discovered those who expressed strong positive emotions while they were in early adulthood lived 10 years longer on average [14]. Another study found that men suffering from high anxiety were **6 times more likely to suffer sudden cardiac death** compared to calmer men [15]. Another study involving 1,200 people that had poor health, found that those who were taught self regulation techniques (the ability to alter unhealthy emotional and mental attitudes) were four times more likely to be alive 13 years later, compared to the control group [16]. This is a major finding because it clearly shows that self-regulation techniques can improve one's health and increase lifespan.

Additionally, studies have proven that an animal's heart-rhythm pattern can shift in response to a human's emotions. This was proven in a collaborative study involving Ellen Gehrke, Ph.D. [17] who used self-regulation techniques to consciously shift into a coherent state while she sat in a corral with her horse with no physical contact. As she began shifting into coherence, her horse's heart rhythm began shifting into an ordered pattern.

Learning to Control Emotions Builds Resilience

Resilience is not something we are born with, it has to be learned much like the qualities of self-love and gratitude are a life-long learning process. Wouldn't it be great if we could fast track our resilience without having to go through all the pain and struggle needed to gradually build it up? Perhaps self-regulation techniques can help. Emotions and Resilience are closely related due to the fact that emotions are primary drivers of major physiological processes that are involved in energy regulation. Resilience is the capacity to prepare for, recover from and finally adapt during adversity, stress, trauma or challenge [18]. Hence one's ability to sustain optimal function, good health and resilience is the key to managing one's emotions. Hence resilience is rather a state of being, rather than a trait. Also a person's resilience varies over time as circumstances, level of maturity and demands change [19].

The 4 Main Types of Resilience

Physical Resilience

This is simply a state of physical endurance, flexibility and strength

Emotional Resilience

This is reflected in one's ability to self-regulate their emotions via adequate emotional flexibility and have a positive outlook with supportive relationships.

Mental Resilience

This is reflected by one's ability to sustain attention and focus with the capacity to integrate multiple points of view.

Spiritual Resilience

This is a commitment to tolerance others' values, intuition and beliefs.

Summary

A person's ability to sustain and build coherence is related to his or her ability to affectivity initiate self-management techniques and deploy efficient utilization of energy resources across the four main domains. These are emotional, physical, mental and spiritual. A high level of resilience is important for bouncing back from a challenging situation as well as to prevent unnecessary stress reactions such as impatience, frustration and anxiety. These wasteful emotions lead to a depletion of one's energy and waste valuable time. They also deplete one's psychological and physical resources.

Self-Regulation and Health

Today more and more health care providers understand that health is not an absence of disease, but exists as a process by which one maintains their state of inner coherence. Coherence in this context being a life that is meaningful, has purpose, is comprehensible and is manageable [20]. Research has proven that when one can self-regulate their emotions (ie have control over anger, frustration and anxiety) that they are able to create major improvements in their self-regulatory capacity. This process works similar to the strengthening of muscles, which in turn makes one less vulnerable to a depletion of their internal reserves, which in turn provides the necessary fuel for resilience [21].

While some people may fail in their ability to learn self-regulatory skills due to trauma, mental impairment or immaturity [22], overall the keys necessary to increase one's ability to self-regulate their emotions, attitudes and behaviours is attributed to the four main traits just covered.

A study involving 5,716 middle-aged participants discovered that the ones with the strongest self-regulation behaviours were 50 times more likely to be **alive without chronic disease 15 years later** compared those that scored low in a self-regulation questionnaire [23].

Anger and Heart Attacks

A University of Harvard Medical School Study examined 1,623 heart attack victims and discovered that the people who had became angry during an emotional conflict had **twice the chance of a heart attack** compared to those that remained calm during an emotional confrontation [24].

It is a proven fact that over one-half of people suffering from heart disease have heart disease that was not caused by the standard risk factors such as high smoking, cholesterol or a sedentary lifestyle [25]. This could explain why the comedian George Burns, who smoked a lot of cigarettes, lived to such a long age. He was a comedian, which is a profession that happens to generate a lot of positive emotions. The opposite would also be true. A person who smoked and suffered from high anxiety and constant negative emotions would be more at risk of dying early compared to a person who did not smoke and exhibited positive emotions, especially during times of crisis.

Personal Mastery and Health

An study looking at 2,829 people in different countries aged between 55 and 85 discovered that the ones who exhibited the highest levels of personal mastery (feelings of control over various life events) exhibited approximately a 60% lower risk compared to those who felt helpless during a life challenge [26]. Another study undertaken by the Mayo Clinic found that people who suffered from heart disease, that psychological stress was the strongest predictor of a person suffering a future cardiac event (cardiac arrest, heart attacks or cardiac death) [27].

Why Constant Feelings of Anger are Unhealthy

Independent studies have found that the risk of developing heart disease increases significantly for people who impulsively vent their anger or for those who repress angry feelings [28] [29]. Hence a healthy balance of 'letting off steam' is healthy, but repetitive impulsive outbursts of anger over a long period of time can be detrimental.

Self-Mastery is more Beneficial than a High IQ

Another study found that qualities such as altruism, self-awareness, motivation, and compassion, especially a person's ability to self-regulate and control their impulses and transform negative emotions, were more important than a high IQ [30]. Hence having the ability to deploy self-regulation skills when necessary enables a person to excel when faced with life's challenges. Hence this could be why some people placed in responsible positions end up ruining a company through false or misleading financial reports or earnings or other forms of manipulation when the going gets tough. This is because some employers in these professions think people with high IQ's will enhance their company's success. However, if the person(s) employed in these positions know how and when to utilize self-regulation techniques during the tough times, they will be a much more valuable asset to the company. In summary, smarter is not always better if the person cannot self-regulate their emotions when in the face of adversity.

The Nervous System

We experience our nervous system mostly through emotions (touch, taste, feelings). The key to self-mastery is learning to control the nervous system. The nervous system is composed of two major parts which has a dual polarity, much like the positive and negative terminals of a battery. Physiologist Walter Bradford Cannon states that when we become nervous, the sympathetic nervous system becomes energized (fight or flight). The opposite part of this is the parasympathetic nervous system, which is the calming part of the nervous system. It slows down our heart rate, calming us down. As I have written extensively about in my Associative Remote Viewing Book series, the nervous system plays a major role in the success of an Associative Remote Viewing Session. When the nervous system is calm and relaxed, which takes place when the body is in mind / heart coherence, the accuracy of an ARV session is greatly enhanced. It may be that a combination of practicing HeartMath self-regulation techniques and inhaling essential oils that contain limonene / linalool create additional coherence synergy.

Changes in the Nervous System Attributed to Cosmic Rays

A study that involved 10 people over the period of 30 days, whose nervous systems were wired up to a monitoring device via skin patches, discovered that their autonomic nervous systems responded to major changes in the cosmic ray density and solar activity. Mild effects were also noted from variations in earth's geomagnetic field [31]. The largest effects on the human body came from the sun's solar wind, the sun's 10.7cm radio flux and the intensity of the Schumann resonance. The study concluded that activity in a person's autonomic nervous system changes when disturbances occur in earth's geomagnetic field and changes in solar activity. The study also concluded that the nervous system is synchronized with earth's geomagnetic field and earth's Schumann resonance. Other studies have confirmed that a relationship exists between solar activity synchronization, geomagnetic activity and the body's physiological rhythms [32 to 36].

The course of evolution has allowed the human body to adapt to changes in the environment. Changes include sudden variations in solar and geomagnetic activity. Back during biblical times, the effects of solar weather on the body may have been much less. This may have been due to a cosmic event in a distant galaxy that impacted earth's ionosphere / magnetosphere, weakening it to such an extent that after the event, above average amounts of solar radiation now leak through earth's ionosphere, with reduced cosmic rays which have negative consequences. If this is true, humans will eventually adapt to this change, and lifespan will gradually increase.

Today sudden, sharp and above average geomagnetic and

solar activity, especially geomagnetic storms, cause stress in some people's physiological systems. The most significant changes are alterations in the body's serotonin and melatonin levels. Other physiological systems affected include: the immune system, blood pressure, neurological processes and reproductive and cardiac systems [37 to 40].

Cycles of Solar Activity

Highs and lows of solar activity occur over a 10.5 to 11 year cycle. It is during the peak of the cycle that the sun emits above average ultraviolet (UV) energy and increased solar radio flux. These energies can be measured by the 2.8 GHz signal (F10.7cm solar radio flux) [41] [42].

Person's Most Susceptible to Above Average Geomagnetic Disturbances

People vary in their ability to cope with Earth's changing magnetic field. Some are able to quickly adapt to subtle changes in these environmental variables [43] including abrupt and sudden changes. Studies have found that age range [20-50] is not important when looking at a person's sensitivity to local magnetic   fields.

If a person is suffering from a disease, and they are what's termed "Solar Weather Sensitive" and a geomagnetic storm suddenly occurs, if they have a cardiac related disease, the sudden and unexpected increase in geomagnetic activity may result in myocardial infarction incidence and death. Other changes that may occur include: changes in blood pressure aggregation, seizures in epileptics, coagulation and cardiac arrhythmias [44 to 55].

Scientists at the Lithuanian University of Health Sciences

recently developed specific methods based on HRV variables that can immediately evaluate a person's sensitivity to Earth's magnetic field (Human heart rhythm sensitivity to earth local magnetic field fluctuations. Abdullah Alabdulgade et al. April 2015). Their studies and methods may soon play important roles in the future for health and healing.

The Vagus Nerve and the Nervous System

Published studies have found that the evolution of the human autonomic nervous system, especially the vagus nerves, (which are situated below the ear lobes) are key to the development of emotional experiences [56]. They also appear to regulate one's ability to self-regulate and a person's social engagement [57]. Heart rhythm coherence generates an increase in the beat-to-beat variability in both blood pressure and heart rate. This is equivalent to increases in the rate of change. This causes increases in the vagal afferent traffic that is sent from the cardiovascular system and heart and to the brain. Generating increases in vagal afferent traffic utilizing a non-invasive method (eg heart/based emotions refocusing or heart rhythm coherence) offers a number of potential benefits. Recently a few clinical applications for increasing vagal afferent traffic have been discovered. Most of these cause an increase in afferent activity by using external or even implanted devices which stimulate the vagal afferent pathways. Some of these target the left vagus nerve. Vagal stimulation happens to be an FDA-approved treatment for treatment of those suffering from epilepsy and it is under investigation to be used as a therapy for depression, anxiety, obesity and Alzheimer's disease (Groves & Brown, 2005 Kosel & Schlaepfer, 2003).

It has been shown that an increase in the levels of vagal traffic reduces pain via pathways that travel from the body to the body's thalamus at the level of the spinal cord. A recent study found that stimulating the vagal pathways caused significant reductions in migraine and cluster headaches (Mauskop, 2005). Stimulating the vagal nerve has also been proven to improve memory and cognitive processing (Hassert, Miyashita, & Williams, 2004).

Now, to close out this fourth edition, let's re-review HRV and its impacts on the body. As stated at the start of this book, it is the heart that receives pre-stimulis responses before the brain, acting as a detector for future events, especially emotional events. These are key elements one should have a thorough understanding of when wanting to perform successful associative remote viewing sessions.

References. Chapter 30.

1. Brotman, D.J., S.H. Golden, and I.S. Wittstein, The cardiovascular toll of stress. Lancet, 2007. 370(9592): p. 1089-100.

2. Marchand, A. and P. Durand, Psychological distress, depression, and burnout: similar contribution of the job demand control and job demand-control-support models? J Occup Environ Med, 2011. 53(2): p. 185-9.

3. Nerurkar, A., et al., When physicians counsel about stress: results of a national study. JAMA Intern Med, 2013. 173(1): p. 76-7.

4. Cummings, N.A. and G.R. VandenBos, The twenty years Kaiser-Permanente experience with psychotherapy and medical utilization: implications for national health policy and national health insurance. Health Policy Q, 1981. 1(2): p. 159-75.

5. Hilton, J., On the Influence of Mechanical and Physiological Rest1863, london: Bell and Daldy.

6. Fredrickson, B.L., Positive emotions, in Handbook of Positive Psychology, C.R. Snyder and S.J. Lopez, Editors. 2002, Oxford University Press: New York. p. 120-134.

7. Isen, A.M., Positive affect, in Handbook of Cognition and Emotion, T. Dalgleish and M. Power, Editors. 1999, John Wiley & Sons: New York. p. 522-539.

8. Wichers, M.C., et al., Evidence that moment-to-moment variation in positive emotions buffer genetic risk for depression: a momentary assessment twin study. Acta Psychiatr Scand, 2007. 115(6): p. 451-7.

9. Fredrickson, B.L., The role of positive emotions in positive psychology. The broaden-and-build theory of positive

emotions. American Psychologist, 2001. 56(3): p. 218-226.

10. Fredrickson, B.L. and T. Joiner, Positive emotions trigger upward spirals toward emotional well-being. Psychological Science, 2002. 13(2): p. 172-175.

11. Fredrickson, B.L., et al., What good are positive emotions in crises? A prospective study of resilience and emotions following the terrorist attacks on the United States on September 11th, 2001. Journal of Personality and Social Psychology, 2003. 84(2): p. 365-376.

12. McCraty, R. and D. Tomasino, Emotional stress, positive emotions, and psychophysiological coherence, in Stress in Health and Disease, B.B. Arnetz and R. Ekman, Editors. 2006, WileyVCH: Weinheim, Germany. p. 342-365.

13. Pressman, S.D., M.W. Gallagher, and S.J. Lopez, Is the emotion-health connection a "first-world problem"? Psychol Sci, 2013. 24(4): p. 544-9.

14. Danner, D.D., D.A. Snowdon, and W.V. Friesen, Positive emotions in early life and longevity: Findings from the nun study. Journal of Personality and Social Psychology, 2001. 80(5): p. 804 -813.

15. Kawachi, I., et al., Prospective study of phobic anxiety and risk of coronary heart disease in men. Circulation, 1994. 89(5): p. 1992-7.

16. Grossarth-Maticek, R. and H.J. Eysenck, Creative motivation behaviour therapy as a prophylactic treatment for cancer and coronary heart disease: Part I--Description of treatment [published erratum appears in Behav Res Ther 1993 May31(4):437] [see comments]. Behav Res Ther, 1991. 29(1): p. 1-16.

17. Ellen Gehrke, Ph.D. who consciously shifted into a coherent state while sitting in a corral with her horse,

18. McCraty, R. and M. Atkinson, Resilence Training Program Reduces Physiological and Psychological Stress in Police Officers. Global Advances in Health and Medicne, 2012. 1(5): p. 44-66.

19. Luthar, S.S., D. Cicchetti, and B. Becker, The construct of resilience: a critical evaluation and guidelines for future work. Child Dev, 2000. 71(3): p. 543-62.

20. Grossarth-Maticek, R. and H.J. Eysenck, Self-regulation and mortality from cancer, coronary heart disease and other causes: A prospective study. Personality and Individual Differences, 1995. 19(6): p. 781-795.

21. Baumeister, R.F., et al., Self-regulation and personality: how interventions increase regulatory success, and how depletion moderates the effects of traits on behaviour. J Pers, 2006. 74(6): p. 1773-801.

22. McCraty, R. and M. Zayas, Cardiac coherence, self-regulation, autonomic stability, and psychosocial well-being. Frontiers in Psychology, 2014. 5(September): p. 1-13.

23. Butler, G.C., B.L. Senn, and J.S. Floras, Influence of atrial natriuretic factor on heart rate variability in normal men. Am J Physiol, 1994. 267(2 Pt 2): p. H500-5.

24. Mittleman, M.A., et al., Triggering of acute myocardial infarction onset by episodes of anger. Determinants of Myocardial Infarction Onset Study Investigators. Circulation, 1995. 92(7):.

25. Rosenman, R.H., The independent roles of diet and serum lipids in the 20th-century rise and decline of coronary heart disease mortality. Integr Physiol Behav Sci, 1993. 28(1): p. 84-98.

26. Penninx, B.W., et al., Effects of social support and personal coping resources on mortality in older age: the

Longitudinal Aging Study Amsterdam. Am J Epidemiol, 1997. 146(6): p. 510-9.

27. Allison, T.G., et al., Medical and economic costs of psychologic distress in patients with coronary artery disease. Mayo Clinic Proceedings, 1995. 70(8): p. 734-742.

28. Siegman, A.W., et al., Dimensions of anger and CHD in men and women: self-ratings versus spouse ratings. J Behav Med, 1998. 21(4): p. 315-36.

29. Carroll, D., et al., Blood pressure reactions to the cold pressor test and the prediction of ischaemic heart disease: data from the Caerphilly Study. Journal of Epidemiology and Community Health, 1998. 52: p. 528-529.

30. Goleman, D., Emotional Intelligence. 1995, New York: Bantam Books.

31. Synchronization of Human Autonomic Nervous System Rhythms with Geomagnetic Activity in Human Subjects. Rollin McCraty et al. July 2017.

32. Wölk, C. and M. Velden, Detection variability within the cardiac cycle: Toward a revision of the 'baroreceptor hypothesis'. Journal of Psychophysiology, 1987. 1: p. 61-65.

33. Wölk, C. and M. Velden, Revision of the baroreceptor hypothesis on the basis of the new cardiac cycle effect, in Psychobiology: Issues and Applications, N.W. Bond and D.A.T. Siddle, Editors. 1989, Elsevier Science Publishers B.V.: North-Holland. p. 371-379.

34. Lane, R.D., et al., Activity in medial prefrontal cortex correlates with vagal component of heart rate variability during emotion. Brain and Cognition, 2001. 47: p. 97-100.

35. McCraty, R., Atkinson, M., Tomasino, D., & Bradley, R. T, The coherent heart: Heart-brain interactions, psychophysiological coherence, and the emergence of

system-wide order. Integral Review, 2009. 5(2): p. 10-115.

36. McCraty, R., M. Atkinson, and R.T. Bradley, Electrophysiological evidence of intuition: Part 2. A system-wide process? J Altern Complement Med, 2004. 10(2): p. 325-

37. Delamater, A.M., et al., Stress and coping in relation to metabolic control of adolescents with type I diabetes. Journal of Developmental Behavioral Pediatrics, 1987. 8: p. 136-140.

38. Goldstein, D.S., Stress, allostatic load, catecholamines, and other neurotransmitters in neurodegenerative diseases. Endocr Regul, 2011. 45(2): p. 91-8.

39. Frese, M., Stress at work and psychosomatic complaints: a causal interpretation. Journal of Applied Psychology, 1985. 70(2): p. 314.

40. Gaines, J. and J. Jermier, Emotional exhaustion in a high stress organization. Academy of Management Journal, 1983. 26(4): p. 567-586.

41. Huang, M., et al., Identification of novel catecholamine containing cells not associated with sympathetic neurons in cardiac muscle. Circulation, 1995. 92(8(Suppl)): p. I-59.

42. Gutkowska, J., et al., Oxytocin is a cardiovascular hormone. Brazilian Journal of Medical and Biological Research, 2000. 33: p. 625-633.

43. McCraty, R., M. Atkinson, and R.T. Bradley, Electrophysiological evidence of intuition: Part 2. A system-wide process? J Altern Complement Med, 2004. 10(2): p. 325-36.

44. Wölk, C. and M. Velden, Detection variability within the cardiac cycle: Toward a revision of the 'baroreceptor hypothesis'. Journal of Psychophysiology, 1987. 1: p. 61-65.

45. Wölk, C. and M. Velden, Revision of the baroreceptor hypothesis on the basis of the new cardiac cycle effect, in

Psychobiology: Issues and Applications, N.W. Bond and D.A.T. Siddle, Editors. 1989, Elsevier Science Publishers B.V.: North-Holland. p. 371-379.

46. Armour, J.A., Anatomy and function of the intrathoracic neurons regulating the mammalian heart, in Reflex Control of the Circulation, I.H. Zucker and J.P. Gilmore, Editors. 1991, CRC Press: Boca Raton. p. 1-37.

47. Armour, J.A. and J.L. Ardell, eds. Neurocardiology. 1994, Oxford University Press: New York.

48. Cameron, O.G., Visceral Sensory Neuroscience: Interception2002, New York: Oxford University Press.

49. Kukanova, B. and B. Mravec, Complex intracardiac nervous system. Bratisl Lek Listy, 2006. 107(3): p. 45-51.

50. Armour, J.A., Peripheral autonomic neuronal interactions in cardiac regulation, in Neurocardiology, J.A. Armour and J.L. Ardell, Editors. 1994, Oxford University Press: New York. p. 219-244.

51. Cantin, M. and J. Genest, The heart as an endocrine gland. Pharmacol Res Commun, 1988. 20 Suppl 3: p. 1-22.

52. Strohle, A., et al., Atrial natriuretic hormone decreases endocrine response to a combined dexamethasone-corticotropinreleasing hormone test. Biol Psychiatry, 1998. 43(5): p. 371-5.

53. Butler, G.C., B.L. Senn, and J.S. Floras, Influence of atrial natriuretic factor on heart rate variability in normal men. Am J Physiol, 1994. 267(2 Pt 2): p. H500-5.

54. Vollmar, A.M., et al., A possible linkage of atrial natriuretic peptide to the immune system. Am J Hypertens, 1990. 3(5 Pt 1): p. 408-11.

55. Telegdy, G., The action of ANP, BNP and related peptides on motivated behaviour in rats. Reviews in the

Neurosciences, 1994. 5(4): p. 309-315.

56. Fear and anxiety take a double hit from vagal nerve stimulation. Michael S Fanselow. June 2013.

57. Vagal Flexibility: A Physiological Predictor of Social Sensitivity. Luma Muhtadie. Dec 2014.

Chapter 31. Understanding the role HRV Plays in the Body

HRV and Resilience
When a person has healthy HRV achieved through self-mastery techniques it allows for much better adaptability to life's challenges. This dramatically increases their resilience [1 to 7].

Reduced HRV
During the 1970's researchers discovered that reduced HRV was predicted in cases of autonomic neuropathy in diabetic patients before their symptoms manifested [8 to 10]. Reduced HRV also was associated with a higher risk of death post-myocardial infarction [11]. HRV gradually declines with a person's age [12]. Hence when using HRV to examine future health issues, age-adjusted values should be incorporated into the evaluating system [12].

Low HRV Levels are Good for Health
HRV levels that are low, when adjusted for age, is a strong and independent predictor of future health problems. This is the case in healthy people and in people suffering from coronary artery disease [13] [14]. This is a major finding because it shows that heart rhythms exhibit changes or fluctuations in their electrical activity before the actual negative health issue occurs. This is akin to Kirlian photography where the future illness will show up in a person's aura weeks or months before the actual disease manifests. It is interesting to note that Kirlian photography, like heartbeats, both utilize electrical fields. We possess powerful proof that the electrical activity of the heart puts out a specific electromagnetic

signature that disease is going to occur, allowing one to take preventative steps and measures before the actual disease manifests or the condition worsens. This also holds true for our ARV sessions. By learning the ARV protocols, one can understand future emotions that are about to occur and learn to interpret these emotions and draw them down on paper, giving a clear picture of the future event.

HRV and Neuro-Cognition

Recent studies found an association exists between higher levels of resting HRV and a person's performance on cognitive tests that require executive functions [15]. Enhancing the coherence of HRV has also been shown to improve cognitive abilities [16 to 19] and be beneficial for health, reducing health-care costs [20 to 26]. Parasympathetic fiber activity becomes stronger when HR is below this intrinsic rate during normal activities and when asleep or at rest. If HR rises above ~100 BPM, the balance then shifts, and sympathetic activity becomes stronger.

73 beats per minute is the average number of heart beats occurring in healthy individuals over a 24 hour period. When HR rises above this it can be an important marker of mortality in a wide spectrum of conditions especially cardiovascular morbidity [27].

A key point here is the relationship between HR and the amount of HRV. As a person's HR begins to increase, less time between heartbeats for variability occurs. Hence HRV starts decreasing. At lower HR's more time occurs between the heartbeats. Hence variability starts increasing. This effect is officially called **cycle length dependence**. It is present in healthy elderly people to some degree, even into advanced age. On the other hand, elderly people with ischemic heart disease or similar pathologies will show less variability as their HR's decrease. Increases in sympathetic activity are the main method used to increase a person's HR above the intrinsic level generated by the SA node. Activation of this branch of the autonomic nervous system along with the activation of the human endocrine system, allows one to respond to stressors, challenges or threats by increasing their mobilization of energy reserves.

Coherence during Geomagnetic Storms

Coherence is greatly hindered during geomagnetic storms. From my personal experience of years of utilizing the HeartMath coherence technique, I have found it much harder to go into coherence during times the geomagnetic activity is at above average levels (Fredericksburg K-indices above 11 or higher). Hence this could be why research studies show a major association between geomagnetic storms and decreased HRV. This would make sense because the HeartMath Quick Coherence Technique utlizes the heart (breathing in and out through the heart). Hence disturbances in earth's geomagnetic field would decrease the rate at which coherence occurs because the human cardiovascular system is impacted by geomagnetic disturbances [28 to 37].

VLF

Several studies have found that during geomagnetic storms, that a ~25% reduction in the VLF (very low frequency) rhythm and that the high frequency (HF) rhythms were not greatly impacted [38 to 41]. This frequency is also strongly associated with increased health risk [42].

References. Chapter 31.

1.McCraty, R., Atkinson, M., Tomasino, D., & Bradley, R. T, The coherent heart: Heart-brain interactions, physiological coherence, and the emergence of system-wide order. Integral Review, 2009. 5(2): p. 10-115.

2. McCraty, R. and M. Zayas, Cardiac coherence, self-regulation, autonomic stability, and psychosocial well-being. Frontiers in Psychology, 2014. 5(September): p. 1-13.

3. McCraty, R., Childre, D, Coherence: Bridging Personal, Social and Global Health. Alternative Therapies in Health and Medicine, 2010. 16(4): p. 10-24.

4. Singer, D.H., High heart rate variability, marker of healthy longevity. Am J Cardiol, 2010. 106 (6): p. 910.

5. Geisler, F.C., et al., Cardiac vagal tone is associated with social engagement and self-regulation. Biol Psychol, 2013. 93(2): p. 279-86.

6. Reynard, A., et al., Heart rate variability as a marker of self-regulation. Appl Psychophysiol Biofeedback, 2011. 36(3): p. 209-15.

7. Segerstrom, S.C. and L.S. Nes, Heart rate variability reflects self-regulatory strength, effort, and fatigue. Psychol Sci, 2007. 18(3): p. 275-81.

8. Braune, H.J. and U. Geisendorfer, Measurement of heart rate variations: influencing factors, normal values and diagnostic impact on diabetic autonomic neuropathy. Diabetes Res Clin Pract, 1995. 29(3): p. 179-87.

9. Vinik, A.I., et al., Diabetic autonomic neuropathy. Diabetes Care, 2003. 26(5): p. 1553-79.

10. Ewing, D., I. Campbell, and B. Clarke, Mortality in diabetic autonomic neuropathy. Lancet, 1976. 1: p. 601-603.

11. Wolf, M.M., et al., Sinus arrhythmia in acute mycardial infarction. Medical Journal of Australia, 1978. 2: p. 52-53.

12. Umetani, K., et al., Twenty-four hour time domain heart rate variability and heart rate: relations to age and gender over nine decades. J Am Coll Cardiol, 1998. 31(3): p. 593-601.

13. Dekker, J.M., et al., Heart rate variability from short electrocardiographic recordings predicts mortality from all causes in middle-aged and elderly men. The Zutphen Study. American Journal of Epidemiology, 1997. 145(10): p. 899-908.

14, Tsuji, H., et al., Reduced heart rate variability and mortality risk in an elderly cohort. The Framingham Heart Study. Circulation, 1994. 90(2): p. 878-883.

15. Thayer, J.F., et al., Heart rate variability, prefrontal neural function, and cognitive performance: the neurovisceral integration perspective on self-regulation, adaptation, and health. Ann Behav Med, 2009. 37(2): p. 141-53.

16. McCraty, R., Atkinson, M., Tomasino, D., & Bradley, R. T, The coherent heart: Heart-brain interactions, psychophysiological coherence, and the emergence of system-wide order. Integral Review, 2009. 5(2): p. 10-115.

17. Lloyd, A., Brett, D., Wesnes, K., Coherence Training Improves Cognitive Functions and Behavior In Children with ADHD. Alternative Therapies in Health and Medicine, 2010. 16(4): p. 34-42.

18. Ginsberg, J.P., Berry, M.E., Powell, D.A., Cardiac Coherence and PTSD in Combat Veterans. Alternative Therapies in Health and Medicine, 2010. 16(4): p. 52-60.

19. Bradley, R.T., et al., Emotion self-regulation, psychophysiological coherence, and test anxiety: results from an experiment using electrophysiological measures. Appl Psychophysiol Biofeedback, 2010. 35(4): p. 261-83.

20. McCraty, R., Childre, D, Coherence: Bridging Personal, Social and Global Health. Alternative Therapies in Health and Medicine, 2010. 16(4): p. 10-24.

21. Lehrer, P.M., et al., Heart rate variability biofeedback increases baroreflex gain and peak expiratory flow. Psychosomatic Medicine, 2003. 65(5): p. 796-805.

22. Bedell, W., Coherence and hearlth care cost - RCA acturial study: A cost-effectivness cohort study Alternative Therapies in Health and Medicine, 2010. 16(4): p. 26-31.

23. Alabdulgader, A., Coherence: A Novel Nonpharmacological Modality for Lowering Blood Pressure in Hypertensive Patients. Global Advances in Health and Medicne, 2012. 1(2): p. 54-62.

24. McCraty, R., et al., New hope for correctional officers: an innovative program for reducing stress and health risks. Appl Psychophysiol Biofeedback, 2009. 34(4): p. 251-72.

25. McCraty, R., M. Atkinson, and D. Tomasino, Impact of a workplace stress reduction program on blood pressure and emotional health in hypertensive employees. J Altern Complement Med, 2003. 9(3): p. 355-69.

26. McCraty, R., et al., The impact of a new emotional self-management program on stress, emotions, heart rate variability, DHEA and cortisol. Integr Physiol Behav Sci, 1998. 33(2): p. 151 -70.

27. Palatini, P., Elevated heart rate as a predictor of increased cardiovascular morbidity. J Hypertens Suppl, 1999. 17(3): p. S3-10.

28. Armour, J.A., Neurocardiology--Anatomical and functional principles2003, Boulder Creek, CA: HeartMath Research Center, HeartMath Institute, Publication No. 03-011.

29. Armour, J.A. and J.L. Ardell, eds. Neurocardiology. 1994, Oxford University Press: New York.

30. Delamater, A.M., et al., Stress and coping in relation to metabolic control of adolescents with type I diabetes. Journal of Developmental Behavioral Pediatrics, 1987. 8: p. 136-140.

31. Goldstein, D.S., Stress, allostatic load, catecholamines, and other neurotransmitters in neurodegenerative diseases. Endocr Regul, 2011. 45(2): p. 91-8.

32. Frese, M., Stress at work and psychosomatic complaints: a causal interpretation. Journal of Applied Psychology, 1985. 70(2): p. 314.

33. Gaines, J. and J. Jermier, Emotional exhaustion in a high stress organization. Academy of Management Journal, 1983. 26(4): p. 567-586.

34. Fowers, B., Perceived control, illness status, stress and adjustment to cardiac illness. Journal of Psychology, 1994. 128(5): p. 567-579.

35. Brotman, D.J., S.H. Golden, and I.S. Wittstein, The cardiovascular toll of stress. Lancet, 2007. 370(9592): p. 1089-100.

36. Marchand, A. and P. Durand, Psychological distress, depression, and burnout: similar contribution of the job demand-control and job demand-control-support models? J Occup Environ Med, 2011. 53(2): p. 185-9.

37. Fredrickson, B.L., Positive emotions, in Handbook of Positive Psychology, C.R. Snyder and S.J. Lopez, Editors. 2002, Oxford University Press: New York. p. 120-134.

38. Frese, M., Stress at work and psychosomatic complaints: a causal interpretation. Journal of Applied Psychology, 1985. 70(2): p. 314.

39. Gaines, J. and J. Jermier, Emotional exhaustion in a high

stress organization. Academy of Management Journal, 1983. 26(4): p. 567-586.

40. Fowers, B., Perceived control, illness status, stress and adjustment to cardiac illness. Journal of Psychology, 1994. 128(5): p. 567-579.

41. Isen, A.M., Positive affect, in Handbook of Cognition and Emotion, T. Dalgleish and M. Power, Editors. 1999, John Wiley & Sons: New York. p. 522-539.

42. Wichers, M.C., et al., Evidence that moment-to-moment variation in positive emotions buffer genetic risk for depression: a momentary assessment twin study. Acta Psychiatr Scand, 2007. 115(6): p. 451-7.

A Personal Closing Note about this Edition:

During the week of the final completion review of this fourth edition, a rare complete rainbow appeared. A similar rainbow also took appeared during the final stages of the book Improve your Remote Viewing Accuracy Techniques using Quantum Microtubules. Rainbows are common in Hawaii, but a full complete rainbow from end to end with such clarity and contrast, such as the one

that appeared towards the end of this book is very rare. In Chinese legends, the rainbow represents yin and yang. In Norse mythology a rainbow represents a celestial bridge.

The Celestial Bridge

The Aymara people of Bolivia called the stars in the Orion Belt the Celestial Bridge. A special star in Orion's westernmost belt is called **Mintaka**. This star is very unique because it astrides earth's celestial equator. In Arabic the star means **'The Shining One'**.

The name Celestial Bridge came from the fact that this star can be seen from both the Northern and Southern Hemispheres making this one of the rare stars that can be seen from anywhere on the face of the earth. This unique feature allows Mintaka to rise due east and set due west, with its length of time in the sky being almost exactly 12 hours.

Earth   Auroras   and   Human   Emotion

As is always the case, at the last minute this book went to the publisher for final review, new and relevant information came to light regarding performing the ARV sessions at midnight. As many of you that have followed my work, as well as the information in this book, ARV sessions for best results are preformed around midnight. The information that came to light regards interesting occurrences that occur in earth's atmosphere at around midnight.

Substorms in Space and at Earth

Substorms are responsible for the polar auroras witnessed at midnight in the auroral zone caused by disturbances in earth's magnetic field. During ordinary solar weather conditions, these auroral arcs are seen as usual glowing forms of light, however during the onset of a substorm, they become intensified, and move with rapid progress in a poleward direction and begin expanding until the entire sky is covered. It is during this time that large magnetic disturbances take place. The electric currents that are associated with the substorm period will come down as low as earth's ionosphere, at a distance of approximately 130 km above earth's surface. In space however things get much more interesting. If a synchronous satellite is in orbit around midnight during this time, when a substorm conditions are present, their on-board detectors will register the arriving of numerous electrons and ions. These energetic ions affect the spacecraft. This energy is so intense, that the electrons end up giving it a negative charge to the power of hundreds or thousands of volts, interfering with normal operations.

As we go further out into outer space, in the plasma sheet, we witness extremely fast flows of plasma which are flowing at the speed of approximately 100-1000 kilometers per second. The plasma particles in this region also above average energies and their magnetic fields are changing rapidly and erratically.

The   Stretching   Effect

During this time the magnetic field lines of the tail begin stretching tail-wards and then release themselves, much in the same way as a stretching and rebounding motion of a slingshot. When these lines bounce back, they begin propelling and energizing the ions and the electrons in the midnight region, at longer than average distances.

The law of nature states that natural phenomena must require input of energy, which then is changed into another form. This same law applies to substorms. Hence, this activity occurs when the IMF (interplanetary magnetic field) heads southwards, which is a period when the interplanetary field lines are more strongly linked to the Earth as more energy is flowing from sun's the solar wind towards the magnetosphere. This also happens to be a period where faster "reconnection" between the IMF and terrestrial field lines, begin "peeling away" the magnetic field lines from earth's day side along with the attached plasma. During this process it becomes attached to the IMF which are then dragged into the magnetotail.

If we were to look at this activity from afar, we would see the magnetospheric field lines becoming parted like combed hair with one group moving towards earth's day side around 12 noon and the other group being pulled back into the lobes of the magnetotail. This increased "peeling away" that occurs near 12 noon causes a shift in the balance. 1- Fewer lines head sunward with more begin attracted towards the magnetotail with the cusp moving towards a field line that is anchored closer to the equator. This shift in energy causes two major things to happen

1 -   Earth's magnetic field becomes weakened around 12 noon. This is because the field lines and plasma are peeled off, which causes the solar wind to move closer to Earth. When this happens, the IMF is "southward", with the magnetosphere "nose" heading further earthwards

2 - More of the magnetic field becomes drawn into the magnetotail, causing the tail lobes expand. This allows for the storing of additional magnetic energy in them. Hence, these expanded lobes are storehouse of tremendous amounts of energy that power the substorm.

After long quiet periods, as the IMF rapidly turns southward this reservoir of tremendous energy that charges up as the magnetic field lines and tail fields intensify and in synchronous orbit become stretched tailwards, in a slingshot-like effect, lasting approximately 40 minutes.

A Release of Energy

It is widely held in the literature the critical event causing the release of this energy is due to the formation of an neutral point that becomes X-shaped, a type of neutral line that extends some distance across the tail (not the neutral point of Dungey's theory). This distance is formed quite close to Earth, at a distance of approximately between 15 and 30 RE.

Magnetic Reconnection

The process of magnetic reconnection begins between when oppositely directed field lines occur south and north of the middle of the plasma sheet. The line on the northern side becomes broken in two separate regions at the neutral line. These parts are then spliced to corresponding parts of lines on the southern side which are divided in two. These reconnected and broken halves of lobe field lines than form two new lines. One on the earthward side, which becomes a stretched terrestrial line that rebounds earthward, much like a slingshot has just been released.

The other side is connected tailwards. Because it is no longer connected to Earth, its energy becomes expelled down the magnetotail. Hence, both the plasma riding on it and the plasma further tailwards) forms a plasma bubble called a "plasmoid". Next the newly-reconnected lines become those of the plasma sheet. As the process draws in magnetic field lines from both of its sides and towards the neutral line it soon reaches the magnetotail. Now the effect of the magnetic fields that are piled up in the lobes, including the energy that is stored within them, acts like a pin on a balloon. As the pinhole gradually allows air to escape, releasing energy stored in the balloon, the neutral line allow field lines (with their plasma) to leave the lobe. This causes a reduction in both the energy and intensity of its associated magnetic field. The law of energy states that if energy in nature is conserved, that will disappear in one form, reappearing in another. Hence electric energy that is consumed by a motor will be converted to kinetic energy of motion. When that motion is stopped by friction, the kinetic energy will turn into heat. Magnetic energy that is drawn from the tail lobes also reappears as energy in different forms. Some of this energy is turned into heat, causing a rise in the velocity and energy of electrons and plasma ions. The plasma that is heated in this process is attached to reconnected field lines as those lines are coming from the tail lobes. There are a few particles that share this energy. Hence the energy these receive may be quite big.

Electrical   Currents

A part of this converted energy drives huge electric currents in a circuit that links the plasma sheet and Earth. These connecting links exist as magnetic field lines that conduct electric extremely well, much like a superconductor. These field lines are made up of electrons and ions. The magnetotail contains large electric currents flowing during all times across the plasma sheet. This flow goes from the dawn edge to the evening edge, closing along the magnetospheric boundary. During substorms parts of this current become diverted earthwards along magnetic field lines. This diversion begins in the morning-side half of the plasma sheet. The currents are withdrawn and flow earthward along the field lines. Next they continue in earth's ionosphere, returning to space along additional field lines towards the evening-side of the tail. The section where the cross-tail current is weakened in the middle of the plasma sheet, appears to be a region actively involved in the substorm and the flow of electric currents along these field lines is possibly key to production of substorm aurora. Substorms are akin to earth's thunderstorms. Energy supplied by moisture in warm, humid air, combined with a rising flow creates updrafts that extends to great heights. This flow of controlled air in the central "updraft" and the formation of rain and even of lightning. Substorms have greater distance and size.

Final Summary

During geomagnetic storms at midnight there is a huge amount of energy taking place. The release of this energy takes place in a slingshot type effect, caused by stretching. During these high energy events, earth's magnetic field is weather at the 12 noon side of earth. As we have shown, ARV sessions and RetroPK effects fail during full moons when geomagnetic energy is above average. This could be attributed to this huge amount of energy.

Much of this energy is contained in the southward IMF. It may be that as this energy fades, entering the sweet spot around midnight, a similar, less energetic slingshot type effect takes place, which enhances coherence. Hence coherent energy is the result / energy of large amounts of previous energy.

As we showed in Chapter 26 a study discovered that RNG deviations depend on IMF polarity which coincided with emotionally significant conditions. Hence the energy from these intense storms may be fueling / affecting emotions. It may be that during the 'sweet spot' phase the fading of intensified energy fuels an intensification of coherence that is

An Introduction to Remote Viewing the FOREX.
Schumann Resonance Coherence Secrets.

generated   by   coherence   techniques.

Thank you for reading part 4 of our remote viewing series. I hope the information herein greatly enhances your intuitive abilities and brings clarity to your remote viewing sessions.

Scott Rauvers
Author

The Solar Institute's Remote Viewing Series of Books

CONSTELLATIONS AND REMOTE VIEWING
Book 1 - *Wormhole Theories, Sunspot Activity and Remote Viewing Stocks*. Topics Covered: Quantum Tunneling, Herbs for Remote Viewing, 13:30LST, The Star Arcturus, Cosmic Rays and Remote Viewing, Air Pressure, The Human Nervous System and Precedent Activity, Frequencies that Enhance the Results of Remote Viewing, Solar and Weather Conditions for Prime Associative Remote Viewing Sessions, Intuitive Biorhythms and Remote Viewing, Magnetic Midnight, the Ophiuchus Constellation, Mayer Waves, Moisture as a Medium for Conveying Information, The Associative Remote Viewing Procedure, Studies Involving Remote Viewing the Markets, Torsion Effects and Time, Magnetic Fields, Paramagnetic Materials, Angular Momentum and the Density of Time and much more!

REMOTE VIEWING HARDWARE AND TECHNOLOGY
Book 2 - *Associative Remote Viewing Technology. Secrets of Precognition and Intuition.* **Topics Covered:**
Emotions as Sensors for Future Stimuli, Associative Remote Viewing and power of Expectation, The Maharishi Effect, Remote Viewing the Future of the Dow Jones, Remote Viewing Electronics / Technology, Dealing with Remote Viewing Interference, Schumann Resonance, Heart Math Coherence and Remote Viewing, Humidity as an Emotional Intensifier, Polarized Light, Finding the Ideal Remote Viewing "Sweet Spot", The Key of Time, The Quarter Moon, Neutrinos and the Nervous System, Tungsten and the Electroweak Force, Hydrocarbons, Barometric Air Pressure and Intuition, Maintaining Strong Brainwaves During Remote Viewing Sessions, Triboluminescence, The Color Yellow, Environmental Radiation and Remote Viewing, Biodynamic Gardening Phases and Remote Viewing, Photoelectrics and much more!

THE QUANTUM REALM AND REMOTE VIEWING

Book 3 – *Improve your Remote Viewing Accuracy Techniques using Quantum Microtubules.* **Topics Covered:** The Quantum Mind, Remote Viewing and Quantum Mechanics, The role Microtubules play in Remote Viewing, **Remote Viewing and Non-locality, The Hypothalamus and Remote Viewing, Gems and Minerals that Enhance Remote Viewing, Quantum Coherence, The Hippocampus, Empathy and Psychic Ability, Substances that Enhance Remote Viewing, Linoleic Acid and Quantum Mechanics, Quantum Photosynthesis, Dopamine and Remote Viewing, Transthyretin, Neurotransmitters and Remote Viewing, Lithium, Monoterpenes, The Signal to Noise Ratio and Remote Viewing, Essential Oils and Quantum Effects, Anesthetics, Taxol, The Pacific Yew Tree, Bacteria, Monoterpenes and Quantum Photosynthesis, Consciousness and Frequency, Meditation, Brainwave Rhythmus and Remote Viewing, Photons, Alternate Timelines and Parallel Universes, The Zero Point Field, The Best Moon Phases for Remote Viewing, Favorable Environments and Conditions for Remote Reviewing and much more!**

You may preview the first 3 chapters of any of these books by visiting:

www.ez3dbiz.com/library.html

Index

Notes

Notes

www.ingramcontent.com/pod-product-compliance
Lightning Source LLC
Chambersburg PA
CBHW072011230526
45468CB00021B/1185